Palgrave Studies in Natural Resource Management

Series Editor
Justin Taberham
London, UK

This series is dedicated to the rapidly growing field of Natural Resource Management (NRM). It aims to bring together academics and professionals from across the sector to debate the future of NRM on a global scale. Contributions from applied, interdisciplinary and cross-sectoral approaches are welcome, including aquatic ecology, natural resources planning and climate change impacts to endangered species, forestry or policy and regulation. The series focuses on the management aspects of NRM, including global approaches and principles, good and less good practice, case study material and cutting edge work in the area.

More information about this series at
http://www.palgrave.com/gp/series/15182

Edward W. Glazier

Tradition-Based Natural Resource Management

Practice and Application
in the Hawaiian Islands

Edward W. Glazier
Wrightsville Beach, NC, USA

Palgrave Studies in Natural Resource Management
ISBN 978-3-030-14841-6   ISBN 978-3-030-14842-3   (eBook)
https://doi.org/10.1007/978-3-030-14842-3

Library of Congress Control Number: 2019933321

Cover illustration: Oliver Kinney

This Palgrave Macmillan imprint is published by the registered company Springer Nature Switzerland AG.
The registered company address is: Gewerbestrasse 11, 6330 Cham, Switzerland

This book is dedicated to Ms. Julia Elizabeth Murray Stevens and to Pouli holoʻokoa ʻana a ka la
—best friends always.

Foreword

As a Native Hawaiian who has directed a federal organization that manages fisheries in Hawaiʻi and the US Pacific Island Territories for nearly forty years, I have met and been moved deeply by many Hawaiian, Samoan, Chamorro, and Refaluwasch fishermen who struggle to keep their ancient traditions alive and to pass that knowledge and way of relating to the natural world to the next generation. I have also worked daily with an array of scientists, versed in the Western way of perceiving the world, boiling down phenomena into mathematical equations, running complex models to understand fishery and environmental data, and searching for the best scientific information available. Edward Glazier, the author of this book, is a social scientist who attempts, as I do, to bridge these two ways of perceiving the world—ancient and modern.

I first met Ed in 2005, when the organization I direct, the Western Pacific Regional Fishery Management Council, contracted him to write the proceedings for a three-part series of workshops on ecosystem-based fisheries management, focused on ecosystem science and management, ecosystem social science, and ecosystem policy. At the time, the Council was restructuring its species-based Fishery Management Plans into place-based Fishery Ecosystem Plans (FEPs). The workshop proceedings were published in 2011 as *Ecosystem-Based Fisheries Management in the Western Pacific* by Wiley-Blackwell with Ed as the editor. In the preface, he notes the following:

> An important outcome of the social science workshop was recognition of the ongoing importance of indigenous fishery practices and traditional and local knowledge of marine resources and ecosystems… The Council's approach to ecosystem-based management involves, among other strategies, adaptive management, emphasis on indigenous forms of resource management, and opportunities for community involvement in the management process across archipelagic sub-regions. There was consensus among workshop participants that this was a valid approach and that it should continue to be emphasized by the Council as it moved forward with the FEPs.

In this current book, *Tradition-Based Natural Resource Management: Practice and Application in the Hawaiian Islands*, Ed delves into the history of colonization that threatened to obliterate indigenous communities in Hawai'i and other Pacific Islands along with the natural resources that they had used and managed for millennia. Fortunately, native people and their ties to the ocean and land are strong, so remnants of these native cultures have not only survived but are in a period of restoration and growth. The *Ho'ohanohano I Nā Kūpuna Puwalu* series, which brought together more than a hundred traditional practitioners from throughout the Hawaiian Islands, is one of many endeavors in recent times to help with this renaissance. The Council, in partnership with other organizations, hosted these and subsequent puwalu (gatherings) to integrate indigenous resource management and community involvement into today's governance and educational systems. Ed was invited to participate in these meetings as an observer, and so his writing reflects not only his academic background as a social scientist but also his having witnessed kūpuna (elders), lawai'a (fishing) and mahi'ai (farming) experts, and their 'ohana (families) sharing knowledge as they passionately sought guidance and wisdom from one another and their ancestors on ways to move forward to ensure their culture thrives.

The State of Hawai'i in 2012 officially recognized the traditional 'Aha Moku system of resource management as a direct result of the many puwalu described in this book and the dedication of those who attended them. This success story reflects one of the Council's many initiatives advocating for native fishing and management rights. Soon after its establishment by Congress in 1976, the Council formed a Fishery Rights of Indigenous People Standing Committee. On its recommendation, the Council commissioned five studies, published in 1989 and 1990, on the legal basis for preferential fishing rights for native peoples in Hawai'i, American Samoa, Guam, and the Northern Mariana Islands. The Council was instrumental in having the reauthorized Magnuson-Stevens Fishery Conservation Management Act acknowledge the native people of Hawai'i and the US Pacific Islands and include development, demonstration, and educational programs to assist them in attaining and retaining traditional fishing and fishery management opportunities. Council staff members, such as former indigenous coordinator Charles Ka'ai'ai, communications officer Sylvia Spalding, and program officer Mark Mitsuyasu, have dedicated countless hours to support indigenous communities and traditions not only through these programs but also through other supporting traditional ecological knowledge and climate change symposia; sea turtle and marine planning workshops; community-based fishery management plans; traditional lunar calendars and videos; student art, photo and essay contests, and lesson plans on traditional knowledge; traditional knowledge research; and outreach work regarding a fishing code of conduct based on the testimony and approval of puwalu participants. The Council also catalyzed the creation of the Traditional Knowledge Committee within the National Marine Educators Association as well as the International Pacific Marine Education Network, which promotes both traditional knowledge and Western fisheries science in classrooms and educational policies.

Other organizations and individuals have worked in other ways to stem the traumatic, intergenerational impact Western colonization has had on native people and indigenous land and ocean resources. In *Tradition-Based Natural Resource Management,* Ed elucidates traditional practices that have survived many and various historical constraints. From the resurgence of non-instrument navigation and traditional voyaging canoes to familiar activities like the baby lua'u and other pa'ina (celebratory feastings), Ed shows that the continuation and reclaiming of indigenous culture occurs on many levels and involves both Native Hawaiians and those who have come to call Hawai'i home. I hope reading this book encourages you to become an agent of change by joining this movement. Seek to learn from kūpuna and expert practitioners in your ahupua'a and moku (traditional district), and then mālama (care) for and enjoy the resources in your locality, with due respect for the ancestors, for contemporary elders, and for generations to come. Imua (onward)!

Honolulu, USA   Kitty M. Simonds
February 2019

Preface

The principal intent of this book is to describe more than a decade of meetings held to facilitate discussion of natural resource management issues among Native Hawaiian elders and cultural experts residing on each of the Main Hawaiian Islands. As an outside observer of each meeting, I was continually struck by the impassioned nature of perspectives on matters of profound significance to participants and the communities they represented. Indeed, as a social scientist with deep interest in native societies generally, and specific research experience in indigenous settings around Alaska and Hawai'i, the meetings presented a remarkable opportunity to witness both the contemporary expression of an age-old Polynesian culture and the challenges of indigenous life in the twenty-first century. For this opportunity, I will always be grateful, and it is my hope that this book will somehow benefit the native people of Hawai'i and other regions in the US and abroad. Although the original intent of the text was to synthesize discussion of natural resource issues of importance to Native Hawaiians with those of other indigenous groups in the U.S. Pacific Islands, I found this task to be overly encompassing for two reasons. The first is my own experience in the Hawaiian Islands. While this is limited to pursuit of an advanced degree at the University of Hawai'i and to a series of fisheries-specific research projects around the islands over the last couple of decades, Native Hawaiian culture and society are of particular interest and that which I have worked hardest to understand. There is, of course, no end to learning or attempting to learn, and sometimes one simply has to proceed with a task and let the journey and people do the teaching along the way. In this regard, I offer my deepest thanks to the many Hawaiians and other island residents who tolerated and encouraged me despite my naivete and haole background. I take full responsibility for any and all mistakes made on the way, including those inadvertently made in the following pages. Second, and more importantly, the actual story of the Hawaiian people and their Polynesian predecessors is a massive account, spanning many thousands of years, involving millions of individuals, and encompassing both striking societal accomplishments and much tribulation and sadness. While references are made to other indigenous societies in Oceania and on the North American continent, it was deemed that full analytical synthesis would merely detract from a profound story-in-itself and the lessons it may provide to policymakers, natural resource specialists, and students of indigenous culture in Hawai'i and elsewhere. The following text is primarily descriptive and straightforward in nature. I have merely attempted to use information from extant historical sources and the words of living individuals to compose a narrative focused on past and ongoing interactions between Native Hawaiians and the natural and social worlds around them. This material provides the essential context needed for readers to appreciate the significance of the many 'aha (meetings) of Native Hawaiians that are the principal subject of the book and the present-day outcome of centuries of evolving tradition. The Hawaiian proverb "I ka wā mua, ka wā ma hope" means "the future is found in the past"—that is, the past must be consulted before moving forward with wisdom. This perspective remains at the heart of Native Hawaiian culture and provides the organizing principle for the chronologically arranged narrative that follows.

Wrightsville Beach, USA   Edward W. Glazier

Acknowledgements

I wish to thank and acknowledge the following persons who graciously assisted in the development of this book: Mr. Charles Kaaiai, Dr. Charles Langlas, Dr. Cody Petterson, Dr. Craig Severance, Ms. Sylvia Spalding, and Ms. Julia Stevens. Contributions were also made by Dr. Adam Ayers, Dr. Courtney Carothers, Mr. Rusty Scalf, and Ms. Elyse Butler. Mahalo nui!

Contents

List of Figures

List of Tables

1

Introduction: Traditional Resource Management and Hoʻokumu (Beginnings)

1.1 Overview

The pursuit and use of wild food resources have shaped the human species—our physical nature, our ability to reason and plan for the future, our capacity to interact with others to achieve important societal goals. Although we have moved into an era in which ready-made foods and sedentary lifestyles are commonplace, the human past is deeply imprinted in our genome. Indeed, if a now conservatively estimated 130,000-year lifespan of *Homo sapiens* (Klein 2009) is proportionately represented by the 24-hour clock, and if the advent of the industrial revolution is seen as marking a new era of human social behavior, then our history as hunters, gatherers, and horticulturists had lasted for 23 hours and 52 minutes of the human day.

In an evolutionary sense then, being human equates fundamentally with successful long-term adaptation to basic environmental challenges and opportunities around the planet. Most of the "human day" has been dedicated to the development of understanding about the natural environment and its resources, and efficient means for pursuing, harvesting, consuming, and effectively managing those resources.

© The Author(s) 2019
E. W. Glazier, *Tradition-Based Natural Resource Management*,
Palgrave Studies in Natural Resource Management,
https://doi.org/10.1007/978-3-030-14842-3_1

Modern societies have developed to their current state only because our forebears successfully adapted to the planet's marine and terrestrial environments over the course of time.

Basic life requirements such as acquisition of food, shelter, and medical services continue to drive human behavior across the planet. Adaptive responses to the shifting environmental conditions that condition food security occur at all levels, from the molecular to the macro-social. Today, individuals in most human societies are dependent on commercial-scale agriculture, and most participate in capitalist- or state-based modes of production. As such, the majority of humans are both removed from the direct production of food resources and are in some way subject to market impacts resulting from broad-scale environmental change such as drought or shifts in the availability of seafood. When pursuit of wild food resources does occur in such societies, it is typically for purposes of commerce or recreation under the governing scrutiny of the state.

At the other end of the spectrum, individuals in a small number of societies located in remote parts of the world continue to subsist primarily, and in some rare cases solely, through pursuit, use, and consumption of living marine and terrestrial resources and the products of rudimentary agriculture that require natural resources of arable soil, soil-based minerals, sun, and air. Fully isolated hunting and gathering societies are increasingly rare, and although a few tribal groups in Brazil, New Guinea, and the Andaman Islands continue to resist sustained contact with the outside world, virtually all have in some way been affected by modern technology and other sources of external change (Anderson 2016).

Members of yet other contemporary societies around the world take part in both ancient and modern ways of living. They participate in various forms of contemporary economic production while supplementing the household economy with foods harvested through hunting, fishing, gathering, and small-scale agriculture. Today, as in the past, such activities often involve strong inter- and intra-familial social relationships in which reciprocal sharing of food, labor, and other resources are typical and critically important for survival. This way of life is common across the globe in the twenty-first century, particularly in rural areas where economic opportunities are limited and relationships between people and traditional use of wild food resources have persisted despite centuries of profound social and economic change. This is true in certain rural areas of the United States, for instance, and it is certainly the case among many Native Hawaiians, American Samoans, Chamorros (Guam), Refaluasch (Northern Mariana Islands), Alaska Natives, American Indians, and other indigenous culture groups in what is now the United States and its territories.

Many active members of indigenous groups around the United States and elsewhere in the world retain a deep interest in their ancient cultures and ways of life while creatively negotiating the many requirements and opportunities of modern lifeways. This dynamic process is a core theme of this book, and, because acquisition of food from land and sea is an essential part of the survival equation, and a pivotally important area of indigenous knowledge past and present, special focus is applied to strategies that continue to ensure food security for those involved.

The pursuit and use of wild foods are beneficial in many ways. For instance, farming, hunting, and fishing require cognitive understanding and metabolic energy, thereby contributing to individual and collective fitness—in keeping with the evolutionary architecture of the human body and mind. Such activities also require cooperative interaction with others and thereby provide opportunities for enhancing social relationships between individuals and families as well as within families.

Hunting, fishing, and gathering also require knowledge of when, where, and how to pursue wild foods. Such ecological and practical knowledge is often communicated across generations, thereby positively reinforcing aspects of family and community life, including customary use of wild foods. Among anthropologists, this form of understanding is generally known as traditional ecological knowledge.

Wild foods are also typically rich in organic nutrients and provide immediate dietary benefits to consumers. Such foods are often also shared, bartered, or customarily exchanged or traded in family and community settings, supporting culturally mediated forms of social and economic interaction. When wild-sourced foods are sold in the commercial marketplace, some portion of the monies so generated is often reinvested into natural resource harvesting activities, contributing to household and community economies and cultures in a mutually reinforcing manner.

Finally, the harvest of wild foods in many cases indirectly facilitates conservation of the natural environment. While hunting and fishing may seem contrary to the goals of many non-indigenous conservationists, such activities are often undertaken in keeping with site-specific customary practices that involve careful attention to effects on local ecosystems inasmuch as such effects may affect potential for food production over time. This book discusses various settings in which subsistence-oriented wild food harvesting traditions and environmental conservation objectives are complementary rather than incompatible.

That indigenous persons in contemporary American societies should continue to regularly pursue and use natural resources of land and sea for purposes of sustenance, while also taking part in modern forms of economic production is the combined outcome of history, modern economic pressures, ongoing interest in acquiring and consuming nutritious wild foods and persistent valuation of cultural identity. Each of these factors clearly holds true for many in contemporary Native Hawaiian society, the group that is the respected focus of this book.

Even today, many Native Hawaiians and other Pacific Islanders prioritize activities that were fundamental to the successful colonization of the most remote archipelago in the Pacific Ocean. Descendants of the first colonists continue to hahai holoholona (hunt animals, such as wild boar) in the uplands and mountains; lawai'a (fish) along the nearshore zone and in the deep sea; harvest fish from loko i'a (fish ponds); gather plant materials in many ecological zones around the islands; and maintain small-scale farming operations that require clean water and rich soil. Such persons thereby perpetuate the various customary activities that have long surrounded the pursuit, cultivation, collection, and use of natural resources across the islands.

Many Native Hawaiians continue to perpetuate important customs and ways of life in the present era: they share food, labor, and other resources in extended family and community settings; communicate knowledge of the natural world across generations; and maintain a deep interest in the long oral and written history of the original Polynesian settlers and successive generations of Hawaiians. While many or most indigenous persons in the islands are now of mixed ancestry and varying religious and philosophical orientations, underlying perspectives regarding the importance of traditional interaction with the natural world are consistent. For instance, as many cultural practitioners will readily communicate, taking only what is needed from the ocean enables the ocean to care for the ʻohana (family). This general ethic has been expressed in a variety of ways and places around the islands for many centuries, with place-specific and continually evolving rules and sanctions implemented to help ensure the ongoing availability of resources. These may or may not coincide with natural resource use regulations established by state and federal government agencies in the islands.

In Hawaiʻi, long-standing naʻauao (wisdom) and ʻike (knowledge) about the natural island world are reiterated in various mele (songs, sayings, chants), moʻolelo (stories), and even hula (a form of expressive dance developed by indigenous Hawaiians). In many culturally active families, these are used as means for guiding one's behavior from childhood. But it should be made clear that Native Hawaiians and members of other societies in the U.S. Pacific Islands are in no way "stuck" in the past. Nor is adherence to indigenous customs, rules, and sanctions universal among all such residents. This is obvious—no culture or society is without individual deviation from normative or customary behavior. Rather, observation of the setting makes clear the capacity of core members of Native Hawaiian society to pūlama (cherish or care for) knowledge gathered by past generations while creatively adapting to conditions in the present.

In fact, it may be said that, in light of the continual influx of non-Polynesians over the past three centuries, many descendants of the original colonists have become experts at nurturing traditional knowledge and wisdom while accommodating or adapting to new sources of change. Some have been successful, others less so. Some contemporary Native Hawaiians are simultaneously perpetuating their culture and succeeding in the modern capitalist system—through various culture-based entrepreneurial ventures, through professional positions in the public and private sectors, through smart investment practices, and through otherwise effective participation in the regional and global economies. Many others regularly struggle in an increasingly challenging economic context.

Although economic success is not universal among Native Hawaiians, success in the modern capitalist system is not universally experienced in any society. In the indigenous settings of Hawai'i, the contemporary socioeconomic situation is a complex interface between historic processes and contemporary values that often reflect those of the past. Successful participation in both traditional lifeways and modern capitalist society is possible, but no mean feat, here. The challenges are indicated in various consistently discouraging measures of socioeconomic and public health status (Office of Hawaiian Affairs 2014; Brown et al. 2009; Kana'iaupuni et al. 2005). Unfortunately, this holds true for all indigenous American populations. In American Samoa, for instance, the household poverty rate was 57.8% at the time of the 2010 Census, far higher than any state, territory, or commonwealth in the nation (United States Government Accountability Office 2014). Yet, key aspects of Fa'a Samoa (the traditional Samoan way) remain vibrant among populations of Samoans across the nation.

In the case of Hawai'i, the challenges have been extensive and persistent. This is, in fact, a significant understatement. In the centuries following Captain Cook's first Hawai'i landing on Kaua'i early in 1778, a succession of foreign explorers, missionaries, military forces, and venture capitalists advanced their interests with limited regard to the well-being of the original inhabitants, and in some cases with the intent of overtly oppressing them. Lands were taken and redistributed; long-standing cultural practices were discouraged; diseases were transmitted, resulting in massive population loss; and in the late nineteenth century, the Hawaiian kingdom was eventually overthrown, and illegally so (U.S. Public Law 103–150 (107 Stat. 1510)).

Such historical events and processes have generated long-term effects among members of the host society. Native Hawaiian individuals and families who engage the contemporary economic system often do so from positions that have been conditioned by historically limited capital, land, and political power. Pursuit and use of natural resources and related customs are typically undertaken in a context of historically limited household income, constrained access to land and sea, and limited legal basis for managing such resources as in centuries past. Significantly, a series of 'aha (conferences) co-convened by the

Western Pacific Regional Fishery Management Council, the Association of Hawaiian Civic Clubs, the Office of Hawaiian Affairs, the State of Hawai'i Office of Planning Coastal Zone Management Program, the Hawai'i Tourism Authority, and Kamehameha Schools, led to the passage of legislation that gives Native Hawaiians and local residents of all ancestries the opportunity to formally advise state agencies on matters relating to place-based management of natural resources across the Hawaiian Islands. This process and outcomes to date are described at length later in this book.

Regardless of challenges of past and present, it is clear to those who have lived in Hawai'i and who have interacted with local residents for some time that the vast majority of Native Hawaiians have never abandoned their own cultural identity. In keeping with the proverb, I ulu no ka lala i ke kumu (the branches grow because of the trunk; that is, without the ancestors we would not be here), many Hawaiians have rather moved forward in time with attention to lessons from the past and readiness to persist and flourish in the present and in years to come.

Collective economic success is difficult to achieve in the present within any society. But in Hawai'i and other settings around the nation, basic challenges have been worsened especially because the land base upon which indigenous cultures originated has been diminished radically and sometimes forcibly over time, thereby limiting opportunities for age-old food gathering and related customary practices and constraining the prospects for economic development in the present. The situation has been the subject of ongoing political struggles by Native Hawaiians, many of whom continually and avidly assert their ability to sustainably use the natural environment and its resources for customary and novel purposes.

The capacity of Native Hawaiians to nurture customary aspects of social life while adapting to modern sources of change runs counter to assertions that the group has somehow purposely invented the past to achieve certain objectives in the present (Keesing 2005). Native Hawaiian scholars such as Trask (2005) argue forcefully that such claims are flawed in various ways. For instance, any "reinvention" of the past logically requires a static historic condition that can be reconstituted. In fact, culture and tradition are widely known to be dynamic social phenomena, and oral and written accounts of Hawaiian cultural history abound with instances in which customs and traditions evolved in response to changing environmental, social, political, and other conditions (Cordy 1981). Thus, it is important to recognize that although certain customs, such as a culture-based system of social controls (kapu) on the exploitation of natural resources has persisted in Hawaiian communities for many centuries, the localized expression of that system has varied and continues to vary in rationale, form, and effect over time and space.

Assertions about the invention of tradition in the context of Hawaiian culture have also failed to account for the actual inter-generational tenacity of certain indigenous forms of belief, knowledge, and activity. All cultures and modes of cultural expression evolve continually, but some are relatively stable. Key elements of Hawaiian cultural life have, in fact, persisted over many generations. This may best be exemplified by the ongoing tenacity of the Hawaiian language itself. Indeed, a contemporary form of Hawaiian is being spoken as a second language among many Native Hawaiians and other residents, and earlier, closely related versions persist among smaller groups of Hawaiian-first language speakers. Thousands of persons today are fluent in either form of the language (NeSmith 2005). Despite centuries of exposure to other languages and a sustained period of suppression by missionaries and public officials, the Hawaiian language and, by extension, the overt, subtle, and extensive cultural meanings it represents, remain very much alive in this second decade of the twenty-first century (Brenzinger and Heinrich 2013; Kana'iaupuni et al. 2017).

In many cases, members of indigenous societies around the world value certain traditional aspects of their culture to such an extent that external attempts at oppression can actually strengthen rather than diminish the culture in question. For example, traditional knowledge of the natural environment is particularly important since it can ensure survival of the individual and family. This clearly was the case in the Hawaiian Islands following first encounters with Europeans. Although foreign missionaries and other agents of change tried to alter their "subjects" during the nineteenth and twentieth centuries, certain beliefs and knowledge were so essential and so cherished by Hawaiians that they could not be appropriated or usurped but were rather made huna (secret) and thereby came closer to the heart of the culture.

Many contemporary Hawaiians continue to value and carefully guard certain forms of knowledge about the island and ocean world around them. Moana (ocean) and 'aina (earth) are often considered la'a (sacred), and certain local customs and traditions associated with land and sea are considered inappropriate for discussion with others. This is not a romantic anthropological notion, but rather a social fact that will quickly confront inquisitive outsiders, all of whom are cautioned to hō'ihi (treat with respect) those who are not outsiders, particularly the elders.

This book reviews a wide variety of topics relating to tradition-based management of natural resources in the Hawaiian Islands. Special attention is given to a particularly long indigenous history in which effective use and management of natural resources have been pivotal to physical and cultural survival—beginning with initial colonization of one of the most remote archipelagos on earth and persisting into more recent eras of social and economic marginalization and cultural renaissance.

As emphasized in the following pages, ever-evolving traditional ecological knowledge and site-specific means for ensuring food security are central to Native Hawaiian lifeways past and present. This is true of indigenous societies across Polynesia and in all Pacific Islands administered by the United States. Although certain beliefs about the natural world and approaches to managing and using its resources are often kept secret by Native islanders, many also believe that tradtional knowledge should be communicated.

Reverence and secrecy about certain topics notwithstanding, Native Hawaiians especially find themselves in the position of having to publicly defend their rights to access, pursue, manage, and use natural resources around a chain of islands that for centuries has been increasingly populated, visited, used, misused, and subject to 'ownership' and governance by non-Hawaiians. Many believe that traditional wisdom must be incorporated into modern resource management policies and processes and, that by so doing, all will benefit, even as population-related pressures on island ecosystems and local societies deepen and expand in the present era. The intent of this text is to document and contextualize a historically rooted process of local and district level consultation that advances this important goal.

1.2   A Brief History of the Peopling of Polynesia and Hawai'i

For many years it was thought that ancient seafarers could have done little to manage their course on the vast Pacific Ocean, that the remote island chains of Polynesia were discovered largely by chance and good fortune. But research conducted over past decades, along with real-time trans-Pacific voyages in traditional Polynesian canoes, have made clear that the ancient mariners did not travel aimlessly. Rather, evidence has made clear the ancients were highly adept navigators and at-sea survivalists who carried vital foods and materials on board and harvested creatures of the sea as their canoes transited island destinations known and as yet unseen (Irwin 1992).

The story begins some 60,000 years ago, when human groups migrated southward from Central Asia into the far reaches of the islands of Southeast Asia to reach New Guinea and Australia. According to Irwin (1992: 24), portions of tropical cyclone-free Australasia functioned as a "voyaging nursery in which maritime technology was able to develop for 50,000 years, and a large safety net to which the first tentative voyages of deep ocean exploration could return." Recent perspectives hold that some mariners traveled more directly into the remote Pacific from mainland East Asia.

In any event, forays across the straits and seas of Australasia were part of an expanding human presence that preceded emergence of a distinct cultural tradition known as Lapita. This tradition arose around 3500 years before present, persisted for some 1500 years, and was characterized especially by seafaring and colonization of new islands—first in Near Oceania and, around 3200 years ago, in regions of Oceania no humans had ever seen.

People of Lapita societies were adept farmers and made good use of the marine environment during their travels and as they colonized new areas. The smaller islands of Near Oceania typically provided limited terrestrial resources relative to the adjacent continents and large islands, and so fishing and collection of shoreline and terrestrial foods during such travels was complemented with what Irwin (2006: 74) calls "a portable economy" comprised of various plants and animals.

In conjunction with expanding navigational skills and ecological knowledge, and persistent motivations to travel, such adaptive strategies enabled people of the Lapita culture to voyage into the increasingly distant reaches of the South Pacific (Clark and Anderson 2014). Ecological or demographic pressures probably did not force geographic expansion of Lapita peoples, and underlying purpose may remain a matter of conjecture until better evidence becomes available.

Expansion of human presence into remote parts of Oceania appears to have accelerated some 2200 years ago. Irwin (2006: 76) and Kirch (2000) share the view that the people colonizing or traveling through these areas may accurately be thought of as the people who would become the original Polynesians. In his treatise on Polynesian voyaging *On the Road of the Winds*, Kirch (2000) states that:

> In short, the branch of Oceanic-speaking peoples whom we designate as Polynesians had their origins in the Eastern Lapita expansion to become distinctly Polynesian during the course of the first millennium B.C., within the archipelagos of western Polynesia. Here, in Tonga and Samoa and their close neighbors…is the immediate Polynesian homeland – what generations of later Polynesians would call, in their myths and traditions, *Hawaiki*. (p. 11)

Although an accurate chronological record of human migration in Oceania is challenged by archaeological dating problems, many scholars believe that settlement of West Polynesia was followed by an extensive pause in subsequent long-distance voyaging. Irwin (2006: 76) discusses this hiatus in relation to a variety of possible constraints, including: (a) the need to develop canoes that could successfully undertake what would be particularly long voyages to East Polynesia; (b) wind patterns that were at this time unsuitable for sailing to the east; (c) rising sea levels during the Holocene, which likely reduced the habitability of certain islands and archipelagos; and (d) ecological limitations and social isolation on the relatively small intermediate islands east of the Andesite Line. Significantly, the Andesite Line parallels the deep oceanic trenches around the Pacific Basin, including those adjacent to Melanesia and Australasia and may be seen as a transition zone between near and remote portions of Oceania.

Despite various constraints, there is extensive evidence for a major radiation of ocean travelers throughout East Polynesia some 1500 years ago (Irwin 2006: 77). Again, consistently accurate dating is problematic across the archipelagos, and the range of dates and nature of the dating debates associated with initial settlement and later colonization of this vast region are not reviewed in any depth here. Suffice to say, however, that long before the Viking expansion into Britain and Continental Europe, Polynesian voyagers had developed the skills, knowledge, technology, and level of social organization needed to reach and colonize remote portions of Oceania, including the most geographically isolated archipelagos on earth.

Various forms of research, including computer simulations (Levinson et al. 1973), ongoing experimental voyaging, such as those of the Hōkūleʻa (see Finney 1994, 2006; Polynesia Voyaging Society 2018), analysis of oral traditions (e.g. Taonui 2006); and synthesis of archaeological findings (e.g. Kirch 2000) have clearly demonstrated that discovery of Hawaiʻi and other remote archipelagos did not occur merely by chance or good fortune. Rather, first discovery of new islands across Oceania was largely the result of purposive exploration, the knowledge and skills for which were developed and refined over millennia, first in Australasia, then during the proto-Polynesian expansion into increasingly remote areas of Oceania, and finally by Polynesians themselves during the most recent centuries of this long period of human exploration and settlement of the Pacific Islands (Irwin 1992, 2006).

Purposive exploration and settlement of Polynesia and other parts of Oceania is advanced by numerous scholars. These include Irwin (1992), who, as summarized by Kawaharada (1999), asserts that accumulated knowledge enabled discovery with limited risk and a high rate of survival among the voyagers (Fig. 1.1):

> This deliberate strategy of exploration, according to Irwin, involved waiting for a reversal in wind direction and sailing in the direction that is normally upwind (i.e. eastward in the Pacific) for as far as it was safe to go given the supplies that were carried on the canoe. The return home westward would be made easy when the wind shifted back to its normal easterly direction. Irwin believes that this [general] strategy is supported by the west to east settlement of the Pacific, from the islands of Southeast Asia and Melanesia to Samoa, Tonga, the Cook Islands, the Society Islands, the Tuamotus, and the Marquesas [and ultimately, Hawaii]. The strategy would have been obvious to anyone familiar with sailing. The tradition of ʻimi fenua (in Hawaiian: ʻimi honua), or "searching for lands," reported from Hiva and other Polynesian islands, supports such a notion of deliberate exploration.

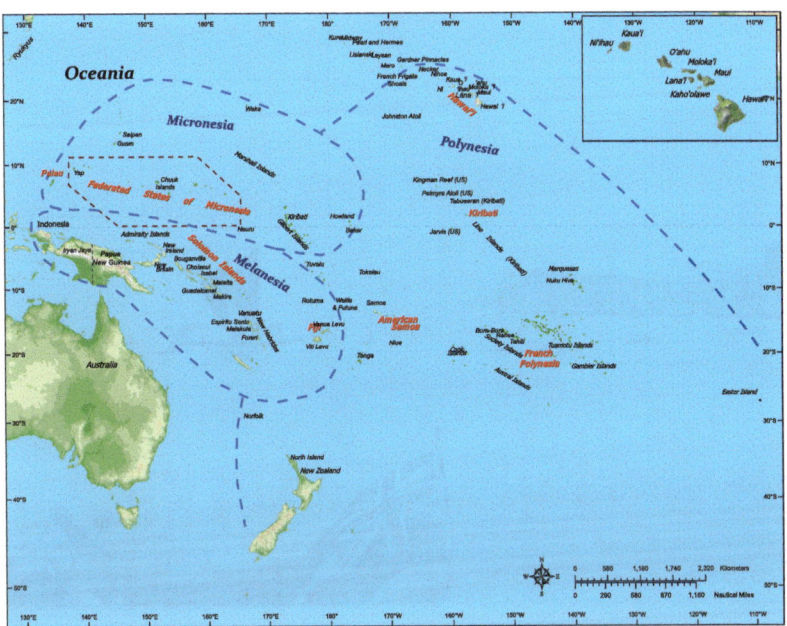

Fig. 1.1 Map of Oceania and Hawai'i

The Polynesian voyagers of antiquity who ultimately reached Hawai'i were more than merely fortunate to have arrived at this remote island chain of eight main islands, now known as: Hawai'i Island, Maui, Lana'i, Kaho'olawe, Moloka'i, O'ahu, Kaua'i, and Ni'ihau. Rather, as made so clear by Finney (2006: 101–152) and Finney and Low (2006: 156–197), these were highly skilled navigators who retained a broad base of evolving traditional knowledge about the Pacific Ocean and its living resources, and about the island environs that were visited and/or colonized along the way. In short, the voyagers ultimately arrived in the Hawaiian archipelago as adept managers of the natural world around them (Fig. 1.2).

Kirch (2010) reports uncertainty regarding the date of first human arrival in the Hawaiian Islands. Based on review of existing evidence, the first canoe probably landed no earlier than around the fourth century A.D. and no later than the tenth century, with the author now tending to support earliest arrival around 800 A.D. based on evidence recovered from the Bellows Dune site on the island of O'ahu. This wide

Fig. 1.2 The Hōkūleʻa returns to Oʻahu after its three-year circumnavigation of earth, June 2017

margin of error relates in large part to improvements in radiocarbon dating technology and sampling techniques, which to support a relatively later peopling of Hawaiʻi from East Polynesia than earlier postulated (Tuggle et al. 1978).

Oral tradition is indefinite regarding the precise place and timing of the arrival of the voyagers and settlement of the Hawaiian Islands (Malo 1951). But an intimate and time-transcendent spiritual relationship between Polynesians and each part of their island world is expressed in many moʻolelo. As Kāne writes in Kawaharada (2004):

The past merges with the present in the telling of some Polynesian legends… Whether something happened a thousand years ago or yesterday makes no difference, the Polynesian is merely the living edge of that great body of ancestral spirits – all the countless lives that have been lived before. The events of their lives are part of his life; he feels that he has participated. (p. 134)

Irrespective of time of arrival, a variety of marine resources sustained the early settlers, as did various endemic plant foods. But of great significance for efforts to colonize new islands, including what would come to be known as the Hawaiian Islands, early and later voyagers also carefully transported a variety of plants, along with animals such as pigs, dogs, and fowl. The process of transporting plants was in itself difficult but critically important, as indicated by Handy et al. (1972):

> The transportation of the Polynesian domesticated plants...was a complicated operation. The crowns of taro, or shoots from the corms, would have to be carefully wrapped to preserve their life, as would also live sweet potato tubers or vine cuttings, banana shoots, root cuttings of breadfruit, and paper mulberry slips. All this required careful planning. (p. 9)

Seafood, endemic plants, and sources of food transported from distant islands supported initial colonization of the Hawaiian Islands. Based on various archaeological findings, Kirch (1985: 287–288) believes that the lush windward sides of the islands were probably settled first, with populations tending to reside along the coast, and with new areas explored and settled over time as the population grew.

Although Hawai'i environs were relatively harsh in comparison with Polynesian archipelagos to the south, many endemic resources were used by the original colonists. According to Bushnell (1993: 7), "about 1,900 species of endemic plants (1,729 represented by seed plants and 168 by ferns) and more than 5,000 species of insects and other small endemic land animals had preceded the human colonists and were thriving in places where water was available." Moreover, the reef, nearshore, benthic, and pelagic environs of Hawaii were yielding of many forms of nutritious marine life. The acquisition of food from the sea was a familiar pursuit and an essential aspect of early subsistence economies among the Polynesians who colonized Hawai'i. Efficient means were developed to utilize a wide variety of species, and new local ecological knowledge was accumulated over time, becoming part of evolving food-gathering traditions.

'Opihi (*Cellana* spp.), papaʻi (crabs), wana (urchins), and limu (seaweeds) were collected by hand along the shoreline, and hukilau (seine) nets, hook and line, and other types of fishing gear were used to capture a wide variety of reef-associated fishes in the nearshore zone. Early Hawaiians also pursued deepwater bottom fish such as ʻōpa-kapaka (*Pristipomoides* spp.), and pelagic species such as aku (skipjack tuna; *Katsuwonus pelamis*) and ʻahi (*Thunnus albacares* and *Thunnus obesus*), among others (Beckley 1883:1-17; Kahāʻulelio 2006:326-330). Fishponds provided a consistent source of protein at some point following the colonization period, as is well documented in the literature. (Apple and Kikuchi 1975: 2; Kirch 1985: 211)

Seafood was complemented with kalo (taro or *Colocasia* spp.), uala (sweet potatoes; *Ipomoea* spp.) and other plant resources gathered or produced across the islands. In conjunction with the tending of agricultural products, fishing and shoreline gathering activities facilitated major expansion of the indigenous population that first colonized Hawaiʻi. Various types of foods became commodities for exchange (Sahlins 1992) and, as discussed in the following section, society became increasingly complex in organizational terms as the centuries passed. Complex chiefdoms approaching state-level societies emerged on the main islands, with the overall population expanding into the many hundreds of thousands of persons on the eve of the arrival of haole (foreign) explorers (Stannard 1989: 45).

1.3   What It Means to 'Manage' Marine Resources in Deep Historic Context

In their recent review of indigenous marine resource management in the Pacific Northwest, Lepofsky and Caldwell (2013) provide instructive discussion of how people of the region passively and/or actively changed environmental conditions to acquire food for family and community. The authors use both archaeological evidence and discussions with living persons to describe relevant physical and cultural processes which, in various ways, can be considered forms of resource management pertinent to analysis of the situation in Polynesia and elsewhere.

These processes include (1) methods used to selectively harvest living marine and terrestrial resources; (2) strategies used to enhance or otherwise alter local ecological systems; (3) establishment of tenure and associated social controls on use of resources; and (4) world views and social relations that influence perception and use of natural resources.

As depicted in Table 1.1, the scheme developed by Lepofsky and Caldwell (ibid.) is useful for envisioning the ways in which indigenous peoples have in the past and, in some settings, continue to use and manage marine resources to meet dietary needs and serve associated cultural interests. Viewed in their interactive totality, these processes may be seen as the essential elements of 'systems' of indigenous resource management, wherein each process *by intent* builds on and interfaces with others in specific sociocultural settings.

As can be noted in the table, human processes can influence the nature of local ecosystems in a variety of ways and through a variety of mechanisms. Each is useful to consider in the context of natural resource management past and present, here with particular attention to the unique nature of human-ecological interactions among ancient and more recent Pacific island societies.

As discussed below in relation to the Pacific islands and specifically the Hawaiian Islands, the development of consistently productive *systems* of natural resource use and management require basic intention and accumulation and use of ecological knowledge at and across each basic phase or process of (a) selective harvesting, (b) ecological enhancement or change, (c) establishment of tenure in specific island districts and smaller parcels of land, and (d) social controls on behavior as these operate in relation to cultural norms, worldviews, and economic imperatives. How these processes conditioned the development of the regional cultures and societies is necessary context for understanding indigenous perspectives on use and management of natural resources now and in the years to come—the core subjects of subsequent chapters of this text. We begin by first reviewing the suitability of island settings for examining systems of natural resource use and management among indigenous peoples, again with directed emphasis on residents of Hawai'i and other archipelagos in the vast Pacific.

Table 1.1 Components of indigenous marine resource management systems, Pacific Northwest[a]

Process	Action	Strategy	Physical evidence
Selective harvesting	Capture desired species	Appropriate capture method Appropriate net mesh size Appropriate timing/level of effort	– Relative size and abundance of recovered taxa
	Capture resources of desired size	Appropriate location	
Ecological enhancement	Increase availability and abundance edible species	Creation of suitable habitat Alter habitat to increase productivity	– Holding ponds – Beaches cleared of stone – Intertidal walls
	Select age and size of edible resources	Transplant eggs or animals Leave spawning animals Leave small animals	– Relative size of recovered taxa
Tenure and social controls	Limit or control access to resources	Establish area use rights Establish ownership of harvest features and/or locations	– Inter-site differences in recovered taxa; Marking of harvest sites; Management features near settlements
	Establish harvest rules	Restrict timing of harvest Establish limits on size of catch Establish harvest eligibility	– Relative size and abundance of recovered taxa – n/a
World view and social relations	Respect for non-human life	Limit harvest to what is needed	– Sustained use of given area – Differential abundance of valued species
	Spiritual connections to animal world	First food ceremonies Return remains to water	– n/a
	Maintenance of kinship ties	Feasting, trading, social events	– Non-local taxa

[a]After Lepofsky and Caldwell (2013)

Islands as Ideal Settings for Understanding Human-Ecological Interactions

It is fitting in this discussion of indigenous resource management in the Pacific islands to begin with a review of the significance of island settings themselves, particularly with regard to their suitability for observation of interactive human-ecological relationships. Various attributes make islands, and perhaps especially small Pacific islands, particularly suitable for considering the role of humans in ecological change (Kirch 1997a: 31) and, conversely, for examining the ways in which ecosystems can be altered to enhance the chances for the long-term survival and well-being of human societies.

Such perspectives are discussed by Vitousek (1995: 11), who asserts that islands afford scientists ideal opportunities for observing the structure and function of marine and terrestrial ecosystems in "relatively simple, well-defined" settings. The resulting understanding can then be used to develop models that are applicable to enhanced understanding of other ecosystems around the globe.

Similarly, Kirch and Hunt (1997) assert that understanding long-term feedback effects of ecological change on Pacific islands may yield much insight into the general functioning of ecosystems in other settings. The following are worth noting: (a) Pacific islands are small relative to continents, distant from large land masses, and diminutive in comparison with the surrounding ocean, all of which lend to observation of ecosystem interactions; (b) bounds between ocean and island and their respective ecological sub-systems are easily envisioned (Berkes 1999: 69) and thereby suitable for study; and (c) marine life tends to aggregate at and around both populated and uninhabited islands, lending to observation of marine ecosystems with and without the effects of direct or indirect human interaction (Sibert and Hampton 2003).

Finally, from a social perspective, the ocean is readily visible from the island setting, which makes it an integral part of daily life. Further, the natural resources contained by the ocean are pivotal in the lives of many residents. Islanders typically perceive living marine resources as finite, vitally important in dietary and cultural terms, often challenging to acquire, and therefore highly valuable. Given that various food-related goods available on continents are not readily attained in the middle of the Pacific, islanders traditionally depend extensively on marine resources, a fact that has long demanded detailed indigenous and local knowledge of marine ecosystems, and the factors and processes that constrain or enable marine resource availability, abundance, distribution, and acquisition (see Poepoe et al. 2003).

Selective Harvesting

Selective harvest of marine and terrestrial resources and the evolutionary implications of choosing certain types and sizes of plants and animals for consumption and other uses are well represented in the archaeological record and literature (Conover and Munch 2002; Swain et al. 2007). Similarities between plants and animals found in new areas and those about which exploring groups were already familiar encouraged harvest and consumption in new regions. Hunger and the need for essential materials undoubtedly were fundamental incentives for travel and migration and for fishing, hunting, and gathering, irrespective of place of origin or destination.

But selective harvesting *intended* by a given human group to improve the capacity of an ecosystem so that it could yield foods or useful materials at rates or volumes greater than might otherwise be possible—a hallmark objective of modern natural resource management—is marked by certain attributes. That is, the group in question would have anticipated long-term occupation of, or return to, a given area, and possessed the knowledge needed to enhance the ecosystem to sustain or improve future harvests. While it can sometimes be determined that a culture group did return to a given island or ocean area to harvest specific foods, it can be relatively difficult to confidently discern the human rationale for selectively harvesting plants or animals based on pertinent attributes, such as size, age, amount of edible flesh, spawning capacity, sex, or other factors.

For example, by selectively harvesting large individuals of a given fish species in a specific nearshore or lagoon ecosystem, moderately sized individuals at optimal spawning age or level of fitness could intentionally be avoided, assuming the harvesters in question retained an interest in conservation for its food production potential and were familiar with the species and its characteristics at-spawning. Although paleo-ichthyological evidence might indicate that spawning individuals of a moderate age or size *were* avoided, this may or may not indicate *intention to enhance resources for future use*. Rather, it could simply be an unintended consequence of the group proceeding with the harvest of relatively large fish based on immediate dietary or economic needs (Hutchings and Fraser 2007). While time-series archaeological data from the same area and regarding human interaction with the same species could help resolve such uncertainty, actual intent is always challenging to infer from the paleo-record. Valid ethnographic information provided by living harvesters or other reliable observers can, of course, aid in the interpretation of intent.

Early Enhancement and Disruption of Pacific Island Ecosystems

Intentional alteration of local environmental conditions can often be reliably interpreted from the archaeological record. This is particularly the case if physical components of the environment were detectably altered in ways clearly indicating an increase in the availability of marine or terrestrial resources or the productivity of harvesting efforts for human use.

Lepofsky and Caldwell (2013) list the presence of holding ponds, stone-cleared beaches, and intertidal walls as archaeological indicators of ecological enhancement among tribal groups in the Pacific Northwest. Given the obvious function of fish holding pens, for instance, it can be readily inferred that construction of pens, weirs, and other physical structures was undertaken to facilitate the capture and/or retainment of harvested marine resources for consumptive use by the tribal group or other society in question.

An excellent example of ecological enhancement of this nature in Oceania is readily available in the many loko iʻa (fish ponds) that remain visible in various states of wear in certain locations on each of the main Hawaiian Islands. Some such ponds were quite large. For instance, the kuapā (wall) of the pond at Heʻeia on Oʻahu measures some 1.3 miles in length overall and 15 feet wide in most portions. Such ponds represented a critically important type of ecological alteration inasmuch as they ensured the availability of a highly nutritious source of food. A wide variety of species were raised and, as described by Kamakau (1976: 47), island areas with many fish ponds were considered "fat"—rich with living marine resources in reserve. Apple and Kikuchi (1975) further indicate the importance of the fish pond in the Hawaiian Islands:

> Practically every culture in the world has practiced aquaculture in some degree … Hawaii had intense true aquaculture. As far as is known, fish ponds existed nowhere else in the Pacific in types and numbers as in prehistoric Hawaii. Only in the Hawaiian Islands was there an intensive effort to utilize practically every body of water, from the seashore to the upland forest, as a source of food either agriculturally or aquaculturally. Fish, crustaceans, shellfish, and seaweed were but some of the products of the total indigenous aquacultural system. Ancient Hawaii's broad aquatic food production system included traps, dams, weirs, and other structures designed to catch mature fish, as well as structures and practices of true aquaculture. (pp. 1–2)

Functional changes to terrestrial zones were equally significant in old Hawaiʻi, as exemplified by extensive terrace systems. Most terracing was undertaken to aid in the cultivation of carbohydrate-rich kalo. The remains of ancient terraces can still be found on both sloping and flat areas across the islands.

Intent is critical when conceptualizing systematic approaches to natural resource management in prehistory, since it suggests that the social actors involved had moved beyond random harvesting and into the realm of trial, error, observation, and accumulation of knowledge. Assuming the actors sought in logical fashion to maximize the use-value of natural resources available to them, their manner of extractive interaction with the natural world would in theory have become increasingly focused and efficient over time.

The position of intent is strengthened particularly in distinct geographic locations where the results of such interactions were consistently observable over time—such as in island settings around the Pacific (Vitousek 1995: 11). That is, consistent habitation of relatively small islands suggests that humans would have had extensive opportunities to learn from observation of outcomes resulting from their own interactions with the surrounding sea and land. By extension, long-term habitation would also suggest accumulation and implementation of functionally useful ecological knowledge—the hard-won basis of any beneficial form of natural resource management strategy.

But quite obviously, many mistakes have been made along the way to gaining ecological knowledge in human-environmental settings throughout the world. Island settings across Oceania are no exception, and early Hawaiians, for example, are thought to have significantly disrupted lowland forests in their working of the land during early periods of colonization (Dye 1994: 5). Moreover, by transporting plants and mammals to Hawaii, the voyagers altered island ecosystems and habitats, in some cases leading to the extinction of certain endemic species (Kirch 1982: 1). Later in Hawaiian history, 'ilihahi (endemic sandalwood trees) were harvested on the main islands, first under limiting kapu established by Kamehameha I but, after his death, with great intensity on the part of the ruling chiefs, incurring various ecosystem and social impacts (Sahlins 1992: 3).

Again, such actions and changes are by no means specific the Polynesia or Hawai'i. In fact, virtually all human societies have in the past affected and continue to affect the natural environment in ways that can be considered at least initially maladaptive. This is the error component of trial and error. Such errors may or may not be corrected over time. For instance, it remains uncertain whether the many human-induced environmental problems observed in this contemporary portion of the Anthropocene (geologic era characterized by inordinate human impacts on the physical environment) are correctable, with implications for the future of our species and the functioning of ecosystems across the planet (Hamilton 2017).

But it must be kept in mind that the functioning of ecosystems in which humans and natural resources are dynamically interactive parts is highly complex, and that the actual effects of human actions may not be readily predicted, accurately observed in the present, or easily detected in hindsight. A good example in the Pacific is the fate of the once-verdant forests and flourishing human populations of Rapa Nui (Easter Island) in Eastern Polynesia.

Notably, Pacific island archaeologist Terry Hunt offers evidence that counters popular and relatively simplistic explanations that colonizing seafarers recklessly deforested the island so that the erection of the massive *marae* (skyward-looking statues) that still line parts of the island could proceed in keeping with the cultural objectives of the day (Diamond 2005). Hunt's (2007) countering evidence regarding the Rapa Nui situation is nuanced and complex, suggesting that deforestation and subsequent decline of the resident human population did not in fact equate with "ecocide" as claimed by Diamond (2005) and others (e.g., Peiser 2005).

Hunt asserts that ecosystem and social impacts were rather the result of interactions among many and various interrelated biophysical and anthropogenic factors and processes. These include but are not limited to (a) the preexisting environmental fragility of this remote singular island; (b) inadvertent introduction by Polynesians of *Ratus exulans* (rats), which wreaked havoc on a wide variety of plants and trees endemic to the Pacific islands; and (c) the effects of salt-laden wind, drought, and erosion on forest ecosystems that were in the midst of being significantly disturbed by rats and somewhat disturbed by humans.

As regards claims that the human population of Rapa Nui diminished drastically as a result of deforestation (Diamond 2005; Peiser 2005; Rainbird 2002), Hunt believes that the situation is better explained by factors that led to population decline elsewhere in the Pacific islands—namely, interactions with newly arriving European explorers who oppressed, enslaved, and brought diseases for which islanders had no immunity (Hunt 2007: 498).

Kirch (1997a: 30–42) uses archaeological findings to examine the long-term responses of two Pacific island societies to various ecological challenges generated by their forebears. For example, the original colonists of the Solomon Islands are known to have generated long-term ecological problems by dramatically altering the forest landscape and extirpating various endemic species—residents of remote Tikopia case-in-point. But later generations of Tikopians addressed the resulting challenges by developing cultural mechanisms for balancing population size with proficient means of production and conservation of natural resources. Some of the cultural mechanisms the Tikopians used for population control, such as infanticide and late-term abortion, are extreme when viewed through our own ethnocentric lens (where ethnocentrism is defined as members of one society failing to envision or appreciate the cultural attributes of another). But survival of the whole population apparently assumed priority, as enforced by chiefly fiat. Draconian measures notwithstanding, equilibrium was reached between population density and food security. The balance so achieved is explained by Kirch (1997a):

> Protein is obtained almost exclusively from the reef and open sea through a sophisticated range of fishing and collecting strategies, the dangers of overexploitation held in check through the exercise of conservation strategies invoked by chiefly sanction (*tapu*)...Let it suffice to say that Tikopia is a model of the sustainable microcosm... (p. 35)

Meanwhile, the response of residents of Mangaia in the Cook Islands to deforestation and related problems generated by *their* ancestors was minimally effective and, in fact, subsequently led to social-ecological problems that could not be surmounted. Disruption of ecosystem productivity (p. 34), violence, and ultimate decline of the human population were the unfortunate outcomes. According to Kirch (1997b: 165), Mangaian society "had [its] back up against an ecological wall marked by scarcity of resources and intense competition [for irrigated field systems and other resources]." Kirch (1997a) writes of Mangaia:

> ...the social terror that pervaded late pre-contact Mangaia was inextricably linked to (I do not say "determined by") the sequence of ecosystem perturbations that had been precipitated [earlier]...The Mangaians were in a very real sense the authors of their history, for in destabilizing and thus biotically impoverishing their island environment, they set up severe constraints that entailed severe cultural responses [after the fact]. (p. 34)

Kirch (1997a: 36) states that prohibition of Tikopian population control mechanisms on the part of Christian missionaries between 1920 and 1950 led to dramatic population growth on the island. This peaked in 1952. The Tikopians struggled to produce sufficient food for the expanding population, and cyclones occurring in 1952 and 1953 caused devastating local impacts in the absence of suitable nutrition. As noted by Glazier (2011), "relief supplies arrived through the intercession of anthropologist Raymond Firth, who was still active in the area after his landmark work with the Tikopians in the 1930s (Firth 1936, 1939, 1967)."

The Tikopian council subsequently monitored population density very closely and some Tikopians were ultimately forced to reside elsewhere in the Solomons. Kirch (1997a) asserts that the Tikopian chiefs had long been acutely aware that the sustainability of ecosystem production in this ocean region "depends upon a delicate balance between human numbers and productive resources" (p. 36).

Establishment of Island Tenure in Ancient Polynesia

A healthy relationship between the size and concomitant demands of a given human population and the availability of food resources is at the core of any effective natural resource management strategy. How the balance was, is, or may be achieved in island settings of the Pacific relates to many factors, not the least of which are the overall carrying capacity of the region in question, cultural strategies developed to ensure food and water security in the near- and long-term, and willingness or ability to move, migrate, or otherwise adapt as needed should conditions change rapidly, as following a cyclone for instance, or over time, as might be the case when soils become gradually less productive or as conditions of drought set in.

Achievement of balance between population and natural and grown resource potential does not need to assume the kinds of population control measures exemplified in ancient Tikopia, although this points to the dire challenges and measures associated with survival in certain island settings. More common responses involved calculated redistribution of people into new areas, prohibitions on entry of new groups into already burdened zones (a cause for warfare in certain instances), development of more efficient means for food production and water collection or distribution, and stringent controls on how, when, and by whom ecosystems are used for harvest or production of food and essential materials.

Early hunter-gatherers on the continents had the option of being nomadic, of following and coming to know the nature of highly nutritious food resources, such as various mega-fauna moving in and migrating across terrestrial ecosystems. Early coastal-oriented hunter-gatherers could also follow and utilize various land mammals along the coastal zone and harvest a variety of living marine resources along the shoreline, in the nearshore zone, and in the open sea. At first glance, expansion of either population of hunter-gatherers was not as conditioned or constrained in environmental terms as in remote island settings, since food-bearing horizons went on relatively indefinitely. Both inland mega-fauna and coastal migration theories are central to contemporary thinking about how the Americas were originally populated (Erlandson et al. 2007; Mandryk et al. 2001).

But in fact, early seafaring Polynesians might also be thought of as capable migratory specialists whose food-bearing horizons went on indefinitely. In this case, however, the migration necessitated a transiting of vast areas of the Pacific Ocean in search of new islands and the sources of food they and the surrounding seas might yield through direct harvest or through basic horticulture and aquaculture. The full range of motives for such intrepid voyages remains uncertain. But the need for sustenance is certain. Although the navigators undoubtedly captured seafood *en route*, island destinations, irrespective of motive for the journey, ideally would bear valuable raw materials, water, fruits, and other foods, if only as needed to replenish stocks for further travel by sea. But as discussed by Holmes (1981):

> Experience had taught the Polynesian that very few edible plants grew on previously uninhabited islands, so with him he took a traveling garden. To Hawai'i he brought about two dozen varieties of plants, though probably not all at the same time. Slips, cuttings, tubers and young plants were first swathed in fresh water-moistened moss, then swaddled in dry

ti-leaf, kapa (bark cloth), or skin from the banana tree. Finally, these bundles were put in lauhala (pandanus leaf) casings and hung from the roof of the canoe's hut. Here they would best be protected from lethal salt water and salt spray. In a few cases, he took seeds.

With such latent food sources in hand, and having found productive soils, water, timber, and other vital natural resources on and around the high islands of Polynesia, some groups stayed and ultimately developed sophisticated means for making the most of the potential dietary contributions of the island and surrounding sea.

The islands of Tonga, Samoa, Tahiti, and the Marquesas held the attention of certain navigators-turned-colonists. Some groups even adapted to the relatively depauperate ecosystems of low-lying atolls (Dickinson 2009), though as noted by Rollett (2002) "our knowledge of human impact on island environments suggests that the smallest islands often suffer the most rapid and acute consequences of human induced environmental change [implying] that resource imbalances among islands likely increased through time." It may well be that it was just such imbalances which inspired long-distance voyaging in the Pacific. In any case, certain groups continued the voyaging tradition, ultimately reaching the northernmost and most distant archipelago of Polynesia—Hawai'i.

There is much evidence for periodic interaction between voyagers across various parts of the Pacific islands even after initial exploration and colonization (Kirch 2000; Rollett 2002; Neich 2006). Having once found new productive lands worthy of long-term habitation, the journeys likely were no longer exploratory, but rather undertaken for purposes of cultural and/or economic interaction (Irwin 1992; Collerson and Weisler 2007). Acquisition of essential material items may have inspired many journeys.

While oral traditions and the work of authors such as Fornander (1878) indicate extensive post-colonization era connections across the Polynesian triangle, clear physical evidence is now also available. Collerson and Weisler's (2007) analysis of stone tools recovered from post-colonization era strata around Polynesia makes clear the range and extent of lingering social connections among island regions. This body of work includes isotopic analysis of lithic materials comprising an adze found on Napuka atoll in the Tuamotus, some 4040 kilometers south-southeast of its point of origin on Kahoʻolawe in the Hawaiian chain. The physical evidence described by the authors makes clear the extent of ongoing social and economic interaction between remote locations across East Polynesia:

> Because [the adze] was collected from Napuka, a low coral atoll in the western Tuamotus in central East Polynesia, the rock from which it was made was transported a minimum distance of 4040 km from its source on Kahoʻolawe in the Hawaiian chain. The likely route between Hawaiʻi and Tahiti via the Tuamotus has favorable winds and currents for two-way voyages. Experimental canoes using non-instrumental navigation made such a journey in 32 days...The Tuamotus, along with the Society Islands, could be approached from all quarters and was thus probably important in Polynesian trade. Our data show that Tuamotu adzes originate from the Marquesas, Pitcairn, Austral, and Society Islands; that is, most of the island groups surrounding the atoll archipelago. Furthermore, because the low coral atolls of the Tuamotus emerged after 1200 CE, and the surrounding island groups were colonized well before then, all imported adzes recovered in the Tuamotus relate to post-colonization interaction with adjacent [and distant] archipelagos. (Collerson and Weisler 2007: 1911)

The historic readiness and capacity of Polynesians to explore and interact across vast stretches of the Pacific might be expected if the region can be seen from the perspective of contemporary Polynesians whose very presence is based on a particularly long heritage of travel by canoe across the largest stretches of open ocean on earth. The Fijian scholar Epeli Hauʻofa (1994, 2005), in fact, subvert notions of isolated Pacific islands, emphasizing rather the ways in which islanders have long been connected by a readily navigable ocean and a common cultural identity as highly adapted "people of the ocean." The Pacific here is envisioned as a traversable sea of islands, a perspective is particularly important in the modern context, as indigenous islanders seek to impart to others the nature of a unique identity that is rooted in ancient ways of negotiating life on and in the Pacific (Hauʻofa 1994):

"Oceania" connotes a sea of islands with their inhabitants. The world of our ancestors was a large sea full of places to explore, to make their homes in, to breed generations of seafarers like themselves. People raised in this environment were at home with the sea. They played in it as soon as they could walk steadily, they worked in it, they fought on it. They developed great skills for navigating their waters, and the spirit to traverse even the few large gaps that separated their island groups. Theirs was a large world in which people and cultures moved and mingled, unhindered by boundaries of the kind directed much later by imperial powers. From one island to another they sailed to trade and to marry, thereby expanding social networks for greater flow of wealth. They traveled to visit relatives in a wide variety of natural and cultural surroundings, to quench their thirst for adventure, and even to fight and dominate. (Hauʻofa 1994)

This perspective is quite similar that of contemporary Native Hawaiian scholar Kanalu G. Terry Young (2012), who discusses Native Hawaiians in relation to their own homeland, which is as much ocean as island:

One specific path for rethinking the Native Hawaiian past is to assert the idea that ʻŌiwi Maoli are the indigenous people of Nā Kai ʻEwalu (the Hawaiian Islands), not simply an earlier arrived immigrant group in Hawaiʻi's contemporary multiracial millieu. Reference to the ʻŌiwi Maoli homeland as Nā Kai ʻEwalu, literally "The Eight Seas" defines islands in a chain by the waters that join them with one another. It is the traditional consciousness that ocean is an extension of island. From shoreline to horizon, ocean is definable as "homeland." It is a familiar realm that guides voyaging canoes by it swells, offers sustenance from its reefs, and endless pleasure through rides upon its waves. (p. 4)

As discussed by Collerson and Weisler (2007: 1911) and other scholars (e.g., Finney 2006; Irwin 1992), at some point in the post-colonization era, long-distance travel across Polynesia began to ebb. Finney reiterates the points made by Irwin (1992), that a series of original exploratory voyages in which the canoe's course was well-documented probably led to return trips to the point of origin and then finally to settlement-oriented voyages to the newly found islands. Ultimately, however, as sufficient people, tools, plants, and animals were available locally, there would no longer have urgent need to undertake long, risky voyages (Finney 2006: 144).

Apparently, voyaging did continue for some centuries for other than utilitarian purposes, with notable examples including Hawaiian pilgrimages to Taputapuātea on Raiatea in Tahiti, and trips made by adventurous Tahitian chiefs seeking to marry into ruling lineages in Hawai'i. But in time, the pilgrimages apparently ceased, as did forays from Tahiti to Hawai'i (Finney 2006: 144; Cachola-Abad 1993; Henry 1928).

Ethnographic accounts of the early European explorers and later missionaries describe extensive within-archipelago travel by canoe, but with a notable lack of discussion about long-distance voyaging. Based on Fornander's calculations, long-distance voyaging ended and Hawaii's isolation from the rest of Polynesia began when La'amaikahiki sailed back to Tahiti around 1450 A.D. (Finney 2006: 143). Notably, Taonui (2006: 45–46) cites Lyons and Alexander (1893), Malo (1903), and Beckwith (1970) in his description of a lengthy series of voyages between Hawai'i and Tahiti, with Kaha'i thought to be the last chief to lead a voyage to Tahiti.

Diminishing involvement in long-distance voyaging is significant as it signals a shift toward the lengthening tenure of extant polities around Polynesia, and a variety of hypotheses have been put forth to explain the change in the years prior to and following first encounters with Europeans (Finney 2006: 143–145). Regarding the potential effect of climate change, Nunn (2000) advances the idea that relatively benign regional weather conditions during the Medieval Warm Period (ca. 750–1200 A.D.) would have facilitated extensive voyaging, but that the cold and stormy conditions characterizing the Little Ice Age (ca. 1400–1850 A.D.) would have constrained long-distance travel. Meanwhile, Caviedes (2001: 234–239) argues that frequent westerlies associated with major El Niño-Southern Oscillation (ENSO) events during the Little Ice Age may have enabled Polynesians to penetrate eastward to Rapa Nui and perhaps as far as South America.

In *Vaka Moana—Voyages of the Ancestors* (Howe 2006), Finney provides a thoroughly contextualized discussion of sailing canoes and their various origins, designs, uses, and meaning in ancient Polynesia. Of particular relevance to the trend of diminishing voyaging, he discounts deforestation as a possible explanation and points rather to regional "priorities," that is, to an increasingly localized focus on social affairs and concurrent social insularity in what had over the centuries become well-established and highly organized Polynesian societies on islands in each of the region's main archipelagos.

Notwithstanding ecological problems on Rapa Nui and other islands, trees of sufficient length and girth for the construction of voyaging canoes were still plentiful on the high islands of Polynesia at the time of contact with Europeans. But based on the accounts of early European explorers, the canoes being constructed in Hawai'i, Tahiti, Fiji, and Aotearoa (New Zealand) at that time suggest within-archipelago travel only, with emphasis on capacity for interisland warfare rather than long-distance voyaging. As noted by Finney (2006: 144), "it is clear that the chiefs were devoting much more of their resources to building large fleets of war canoes and deploying them against their chiefly rivals than to building voyaging canoes and sailing far beyond their shores." For instance, of the nature of Hawaiian canoes during Captain James Cook's arrival in 1778, Finney (2006) uses the description offered by Cook (in Beaglehole 1967):

> As the British were slowly sailing along the Hawai'i Island's southwestern coast, they were quickly surrounded by upwards of 1,000 canoes, among which were 150 large sailing canoes many of which contained 30 and 40 men. However, these were combinations sailing and paddling craft which…were adapted for local travel and trade within the Hawaiian chain, not for distant voyaging. They were also readily converted into the war canoes that Hawaiian chiefs deployed in great numbers to engage in sea battles and invade other chiefdoms. The rounded bottoms of the craft were shaped for ease of handling in Hawaii's rough seas and in landings through the surf, rather than for resisting leeway on long crossings. (p. 144)

Tonga is an intriguing exception to the ultimate cessation of long-distance voyaging in Polynesia. Its well-organized societies appear to have been expanding rather than contracting the voyaging tradition prior to first interactions with Europeans (Finney 2006: 150). The principal motives appear to have been trade and warfare with distant Samoa and Fiji. Given their own lack of forest resources, the Tongans used trees from Fiji and at times the services of expert Samoan canoe-builders to emerge "as a powerful and united maritime chiefdom that extended its military power, political control, and cultural influence to neighboring islands in the Lau Group of southeastern Fiji, to Tafahi, Niuafo'u and Niuatoputapu on the way to Samoa, to Futuna and 'Uvea between Samoa and Fiji, and, more tenuously, to Rotuma, located 250 nm northwest of Fiji" (Finney 2006: 150). Indicative of widespread cultural influences across Oceania, canoes used in the archipelagos of Samoa, Fiji, and Tonga were eventually influenced by canoe and rigging designs used in the ocean and island settings of Micronesia and Melanesia.

With the exception of Tonga, the high island societies of Polynesia appear to have begun to focus their energies inward prior to the fifteenth century A.D. At that point in time, the Hawaiians clearly were in the midst of deepening their knowledge, use, and management of land and sea, with increasingly less external influence and with an indigenous population that was growing in size and social complexity across the main islands.

Social Control as Means for Natural Resource Management in Ancient Polynesia

Abundant water, rich volcanic soil, productive nearshore and offshore resources, and the ingenuity to maximize the food value of island ecosystems underlay ancient Polynesians' successful colonization of the various high island archipelagos. But as noted above, success as indicated by population growth occurred gradually, from (a) an initial period during which founding populations stabilized their presence, especially in verdant windward coastal areas on the various islands; to (b) gradual adaptation to environmental conditions and the effects of their own actions on island ecosystems in various island areas; and finally to (c) development of complex social-organizational polities enabled by increasing efficiency of ecological interaction and population equilibrium. Significantly, at some point(s) in this process, environmental knowledge became increasingly formalized and strictures on use of the natural surroundings were developed and enforced by the chiefs and their konohiki (headmen, resource experts, administrators of rules) with the apparent intent of ensuring long-term ecosystem productivity.

With respect to Hawai'i, there is much debate about the timeline of population expansion, particularly with regard to the size of the indigenous population when Europeans were first encountered. The work of Dye (1994) is now fairly dated but provides a population growth curve and associated discussion that retains much heuristic value.

The author describes two models of Hawaiian population growth before interaction with Europeans. An "arrested growth" model holds that the archipelago-wide population peaked at between 200,000 and 250,000 persons in the seventeenth century, thereafter remaining fairly constant until first encounters with Europeans, with growth constrained by factors including pressures on finite natural resources, climatic changes, and political struggles, i.e., warfare.

This contrasts with a "constant" population growth model wherein island-specific and overall populations continued to grow until interaction with Europeans and the various problems they carried to the islands, especially disease. Dye claims that the constant growth model "contains the potential for larger populations... and is used to support an estimate of between 800,000 and 1,000,000 persons in 1778," as asserted by Stannard (1989).

Like other scholars of the Polynesian past, Dye (ibid.) explores the notion that the ability of Hawaiians to transform a rugged landscape into one that consistently produced sufficient food and vital materials to allow for significant population expansion was in part the result of enforced controls on how land and sea should or could be used. Although various scholars see the world of ancient Hawai'i in significantly different ways (Fontaine 2012: 22), writers such as Cordy (2000: 114) hold that the earliest Hawaiian societies were organized much as they had been throughout Polynesian for centuries prior; that is, in relation to kinship and tenure and with authority based on patriarchal seniority within the kin group.

For scholars such as Hommon (1976, 2013), the ali'i (chiefs) of the early period were not yet substantially differentiated from the kin group, though for others they clearly *were* and assumed positions as intermediaries to the gods or retained god-like attributes themselves (Cachola-Abad 1993). In any event, basic knowledge and guidance about how best to use land and sea to the near- and long-term advantage of localized kin groups presumably were being developed and tested in Hawai'i at this time. This human-ecological focus corresponds with the "founding" or initial phase of colonization in Dye's schema of population change in pre-contact Hawai'i.

A ten-fold increase in population is thought to have occurred during what Dye (1994: 3) terms the "Growth Phase"—a period that, based on various archaeological indicators, lasted from ca. 1150 to 1450 A.D. While it is difficult to determine the precise sequence of events and processes that underlay this period of rapid growth, the author (ibid.: 4) argues that it was an outcome of a well-ordered society that had long drawn upon ecological knowledge garnered both in the Hawaiian Islands and during its long period of exploration and migration across the Pacific. Of significance for this and later generations of inhabitants, success also required close human relationships with the gods (Fig. 1.3):

> [This] model of Hawaiian population trends carries with it implications for other processes at work in old Hawaii. Foremost among these, from an archaeological point of view, is the phase of transforming the natural environment of the islands into a cultural landscape whose elements were assembled and put into place during the long course of Polynesian voyaging and migration.

> This Polynesian cultural landscape, dominated by an orderly system of agricultural fields centered on the production of starchy plant foods such as taro, sweet potato, breadfruit, banana, and yams, was itself part of a well-ordered cosmos where the good works of society both made possible, and were made possible by, the good works of the gods. Over time the landscape of agricultural fields and small settlements, presumably each with its own temple, was created at the expense of lowland forests

Fig. 1.3 Modern hoʻokupu (offerings) at Puʻu o Mahuku Heiau overlooking Waialua Moku, Oʻahu

rich in endemic species…Common sense indicates that evidence should reveal dependent patterns, with forests retreating in the face of agricultural expansion accompanied by increased levels of effort expended in heiau (temple) construction, and that the onset of major changes should coincide with the beginning of the growth phase, when a growing population provided both labor for heiau construction and increased demand for agricultural products. (Dye 1994: 3)

While growing attention to divine powers may have indirectly enhanced food production and population growth, there may also have been a clearly human catalyst for the shift from relatively small nucleated populations to the large, far more complex society that ultimately emerged in old Hawaiʻi. Evidence points to powerful influences from elsewhere in Polynesia.

Dye (1994: 9) and other writers (Kamakau 1992; Malo 1951; Fornander 1878) describe the influential arrival of Paʻao, a powerful chief from either Samoa or Tahiti, who is said to have brought a new pantheon of gods and emphasis on elevated status of the aliʻi around the eleventh century A.D. (Kalakaua 1990: 98). Other scholars are less specific about the source of change and refer more broadly to migrations and influences from Polynesian points south, such as the Marquesas, Samoa, or Tahiti, in the timeframe 1000–1300 A.D. In any event, political changes were evident during this period of rapid growth.

The Convergence of Social Control and Worldview in the Human Ecology of Early Hawai'i

The work of Fontaine (2012) is revelatory in its discussion of varying evidence and perspectives regarding Polynesian migrations to Hawai'i after the first century A.D. and before long-distance voyaging diminished across Polynesia around the fourteenth century. The author concedes that while most traditional indigenous scholars believe that a wave of post-colonization era migrations had occurred, perhaps from Tahiti, perspectives on the nature of preceding and subsequent events and implications diverge from there. Fontaine (2012) also provides the perspective of anthropological scholars who emphasize societal change resulting from internal rather than external agents and processes.

One school of thought holds that the in-migrating Polynesians imposed a stricter and more violent religious and social system than had previously been the case. This shift apparently involved a greater separation between ali'i and maka'āinana (people of the land), with some writers believing that "this increased separation was imposed as a response to societal disorder [extant prior to the arrival of immigrants from the south]" (Fontaine 2012: 29).

Others hold that distinctiveness between commoner and ali'i existed all along as part of a generalized Polynesian world view, in which the ali'i were closely linked to divine realms by virtue of their capacity to recall their mo'okū'auhau (genealogical lineages) from godly ancestors (Cachola-Abad 2000: 79–80). This afforded the ali'i heightened social status, a greater share of mana (spiritual power), and also the responsibility to maintain pono (goodness, balance, righteous order) in social and social-ecological relationships in the island setting. Geneaology and the capacity to trace one's lineage remain important aspects of Hawaiian culture to this day.

Finally, a third school of thought (e.g., Handy et al. 1972: 77–78; Malo 1951: 53) holds that prior to the new migrations, Hawaiians were not led as island societies by aliʻi. Rather, a relatively small population of nuclear and extended families were led by haku (familial overseers), and it was this relatively straightforward system that was ultimately supplanted by a more dichotomous aliʻi-makaʻanaina cosmology following the influx of immigrants from elsewhere in Polynesia.

In all cases, new or renewed emphasis on distinctions between commoner and aliʻi appear to have occurred in conjunction with a wave of migrations from the south, and with this shift came heightened attention and adherence to a system of kapu that by all evidence was intended to maintain pono (well-balanced) relationships between the divine, the aliʻi, the makaʻāinana, and the āina, moana, and lani (sky). Of contemporary significance, Fontaine (2012: 280) states that "many Hawaiians today accept the idea that an earlier and simpler society existed in ancient times that was changed by aliʻi strangers arriving from the south in a second migration."

The fundamental purpose of kapu is to ensure the unimpeded flow of mana in a Hawaiian universe that was or is not comprised of separate realms of heaven and earth but is rather immanent, with all essential elements present at any given moment. Andrade (2008) expresses this perspective in relation to the aliʻi, who seem as intermediaries between the divine and the human, melding all into a harmonious state of being in the Hawaiian present:

> Traditional Hawaiian perceptions of the world do not separate the supernatural from the natural. The Hawaiian world is a seamless place, in which some leaders are often looked upon or given status as godly beings, dwelling among the people and caring for them, as is appropriate for those gifted with more mana, as sign and manifestation of spiritual as well as physical and mental power. (p. 22).

But much was expected of the aliʻi, whose duty it was to ensure that the natural state of pono or universal harmony could be maintained (Fontaine 2012: 31 cites Kane 1997: 35). That is, in order to ensure the flow of mana into the human realm, the ali'i needed to "behave properly, perform rituals correctly, and care for those in his charge." Moreover, according to Fontaine (2012) and other indigenous scholars:

> ...for mana to flow uninterrupted into the human realm, it needed to be separated and protected from defiling elements. This separation was kapu (Barrere 1961; Cachola-Abad 2000:8, 111-118; Kame'eleihiwa 1992:36-37; Kane 1997:26). When kapu was observed, mana could flow unimpeded, harmony prevailed, and in this righteousness the life of the land was perpetuated. When the boundaries maintained by kapu failed, the flow of mana into human society was disrupted, and things began to fly apart. (p. 31)

With respect to scholars whose perspectives hold that internal processes rather than exogenous sources led to significant societal change in Hawai'i after about the tenth century A.D., Fontaine (2012) turns especially to the work of archaeologists such as Cordy (2000), Earle (1978, 1997), Hommon (1976), Kirch (1985, 2000), Sahlins (1958), and Kirch and Sahlins (1992), noting that "although varying in details and presentation, the anthropologist's take on Hawaii's history before European arrivals is consistent." In summary (Fontaine 2012):

> Societal changes came not as a result of outside influences from a migration, but internally. The small groups of chiefs gathered power that they wielded for their own benefit at the expense of the commoners: this is the governing dynamic that the anthropologists see operating in ancient Hawaii. As a result, the close kinship relationships that had characterized Hawaii's people in the first few centuries after contact was radically changed as chiefs became the ruling class, extracting what they needed from the exploited class of commoners. (pp. 81–82)

Whether the result of worldviews that increasingly held the ali'i as sacred intermediaries between the gods and the maka'āinana; or power-based motivations that led to differentiated chiefs and commoners; or some combination thereof, a system of kapu was developed and enforced to maintain order in the ancient Hawaiian universe. Such kapu extended into all realms of life in old Hawai'i, most definitely including human interaction with the natural world, including interactions in the age-old system of use of land and sea described in the following section. Kapu is an important subject of Chapter 3, as expressed in various ways and places prior to and following first encounters with Europeans and even after official abolition of the kapu system subsequent to the death of Kamehameha 1 in 1819.

1.4   Emergence of a 'System' of Natural Resource Management in the Hawaiian Islands

While human-generated environmental problems such as inadvertent loss of certain species and dramatic changes to the landscape occurred in various places around the Hawaiian Islands over the centuries (Kirch 1985), agriculture and fishing-related activities were sufficiently productive over time to enable flourishing populations on the main islands. Struggles and inadvertently problematic ecological interactions notwithstanding, the Hawaiians ultimately became highly successful colonists of a particularly remote and rugged landscape and its challenging ocean surroundings.

The success of Hawai'i's original colonists can be attributed not only to ecological knowledge carried from distant islands but also to learning by trial and error (science) and the gradual development of a hierarchical social "system" that encouraged efficient harvest and production of marine and terrestrial foods. This system did not have as its goal "conservation" of natural resources as commonly defined in the present era—that is, conservation to 'preserve' species and habitats for their own sake or for other-than food-related human interests. Rather, the natural environment and its cultivated products were carefully "managed" to enhance production of food for purposes of survival among small nucleated groups of settlers, and ultimately for sustenance of a rapidly growing and eventually stabilizing population of Hawaiians. This perspective is offered by Kirch (1982: 11), who uses the same logic as Anderson (1979) regarding use of shellfish by the Māori in Aotearoa:

it would be invidious to suggest that the prehistoric Hawaiians were ignorant or unsympathetic to the needs or importance of conservation. Yet, given their burgeoning population and technological limits, conservation "may well have been a luxury they could simply not afford. (p. 64)

Central to the food production system in the Hawaiian Islands of antiquity, and of relevance to ecosystem-level management of marine and terrestrial resources in the islands, was the gradual development of a distinct socio-politically based arrangement for use of land and sea. That is, as populations of colonists grew over time, tenure and place-specific means for using and managing resources began to be established, and these followed the natural contours, resources, and resource potential of the islands.

Maly and Maly (2003 II: 15) assert that by about 1525 A.D., the mokupuni (islands) were politically subdivided into moku (island districts), district sub-portions known as ʻokana or kalana, and smaller land units known as ahupuaʻa (defined in detail below). Yet smaller units included ʻili, kōʻele, māla, and kīhāpai (Maly and Maly 2003 II: 5). Jokiel et al. (2011: 2) assert that ʻili "represented the true basic unit of land division to which [Hawaiians] retained fidelity over long periods of time." As elaborated in various sections of this book, moku, ahupuaʻa, and ʻili are of particular significance in the context of localized use and management of natural resources in Hawaiʻi past and present, with notable variation in historic and contemporary perspectives and experiences between islands and island districts.

Fontaine (2012: 2) describes two distinct scholarly perspectives regarding the emergence of the ahupuaʻa system of land use (ca. 1450 A.D.). As indicated in the previous section, one perspective, expressed primarily by anthropologists (archaeologists), holds that kin-based relationships shifted as power increasingly accrued to an elite class of chiefs who used the fruits of the workers' labor to pursue a privileged lifestyle and various political goals. The other view, held largely by indigenous scholars, many of whom emphasize the importance of oral tradition, holds that chiefs and commoners are interactive in more mutually beneficial ways, with various mechanisms moving the society toward a state of pono. The topic is contested, and the author urges greater integration of the respective disciplines and perspectives with the goal of enriching discussions of ancient Hawaiian society.

The etymology of the term ahupua'a relates to a cairn-like symbol that was used to mark the political-geographic boundaries of distinct sub-units of any given island district; that is, an ahu (shrine comprised of piled stones) surmounted by the image of a pua'a (pig). Ahupua'a can be defined in physical terms as land units that tend to be bounded by the Hawaiian island landscape: characteristically steep mountains at center, uplands leveling out toward the coastline, often between ridgelines of sharp relief; volcanic soils in the lowlands; rocky and sandy shorelines; resource-rich nearshore reef areas; and deep ocean waters with pelagic and benthic resources occurring relatively close to land.

Thus, ahupua'a typically assume a wedge shape, with the narrow point in the mountains broadening along the coast into the broad off-shore realms. That is, the divided lands usually extend from the mountain tops seaward through fertile valleys and on to the outer edges of the nearshore reef (Kamehameha Schools 1994: vi) and beyond. But as noted by Waihona 'Āina Corporation (2006), variability in island terrain influenced the nature of the ahupua'a on certain islands:

> O'ahu and Kaua'i lands tend to go from ridge to ridge and from sea to mountain top. Maui lands tend to go from river to river and sea to mountain, with some ahupua'a having no sea access. Hawai'i lands tend to be long mauka/makai [toward ocean/toward sea] strips, particularly along the Kona coast and there are lands which are exclusively in the mountains.

The ahupua'a can also be defined in political terms, viz., the formal means by which land was (and in some areas continues to be) conceptually if not politically "divided" for use and management by residents and leaders. In terms of production of food and acquisition of vital materials, most ahupua'a are well-bounded valleys, wherein plants, trees, and other resources are tended in the uplands, agricultural products are cultivated in terraced mid-lands and in the low-lands, and marine resources are harvested along the shoreline, nearshore, benthic, and pelagic zones (Kirch 1985: 208; Goto 1986: 448).

Streams, waterfalls, and natural ponds were and are of obvious importance where and when available, though certain (typically leeward) parts of the various islands are often relatively dry. A plethora of ecological products are available in most well-attributed ahupua'a, as was expressed by Lyons (1875):

> The ahupua'a ran from the sea to the mountain, theoretically. That is to say, the central idea of the Hawaiian division of land was emphatically central, or rather radial. Hawaiian life vibrated from *uka*, mountain, whence came wood, *kapa* [bark cloth] for clothing, *olona* for fish line, ti [kī] leaf for wrapping paper, i.e., for rattan lashing, wild birds for food, to the kai whence came 'ia, fish, and all connected therewith. Mauka (landward) and makai (seaward) are therefore fundamental ideas to the native of an island. Land...was divided accordingly. (p. 104)

According to Sahlins (1992), as Hawaiian societies developed in political complexity over the centuries, the labor of commoners was offered in tribute to the ali'i and gods, or god-like ali'i. This thereby reified various kapu on the actions of the polity, including those associated with use of natural resources, such as allowable season or timing of use, type or size of resource that could be used, who could harvest or consume a given resource, and so forth. Again, these are elaborated in Chapter 3.

The great Native Hawaiian historian Samuel Kamakau (1815–1876) provides a first-hand account of how the kama'āina (children of the land; also hoa'āina or native inhabitants) enabled the functioning of a hierarchical political economy through work, skill, and knowledge of land and sea, all in the context of the 'ohana holo'oko'a (extended family) and the gods that provided and interpenetrated life and society (Kamakau 1992):

> The Hawaiians were in the old days a strong and hard-working people skilled in crafts and possessed of much learning. In hospitality and kindness they excelled other peoples of the Pacific...With their hands alone, assisted by tools made of hardwood from the mountains and by stone adzes, they tilled large fields and raised taro, sweet potatoes, yams, bananas, sugar cane, and 'awa [kava]; and bartered (ku'ai) their product

or used it at home. Always the first food of the harvest was offered to the gods. Parents before they died instructed their children, the sons to plant and fish, the daughters to make and dye tapa and weave mats. The land was fertile, and the principal crop on Kauai, Oahu, and Molokai was wetland taro cultivated in ponds, artificially constructed patches, along the banks of water courses, or anywhere where the ground was soft and moist. On Maui and Hawaii where there was less wet land, dry-land taro was cultivated. On Lanai and Niihau sweet potatoes were the principal crop. On Kauai, Oahu, and Molokai also are to be seen most of the fishponds built to preserve the fish supply; very few occur on the other islands. (p. 237)

It was typically the case that people living or working in the forested upland areas of a given ahupua'a shared services and goods with people in the coastal portions, and vice versa. Various residents specialized in the knowledge of upland, shoreline, or offshore resources and cooperated to effectively manage and use those resources within the various ahupua'a and across moku on a given island. The overall food security and economic status of any given moku was therefore defined by the collective attributes of the 'ohana residing in a given ahupua'a. This subsequently influenced the size of the population. This is noted by Sahlins (1992: 20), who uses the example of Kona moku on Hawai'i Island, the ali'i of which utilized the products of ahupua'a encompassing upland and lowland field systems, fertile stream-fed valleys and peripheral fishing zones, and sweet potato fields. The verdant central valleys in this region were relatively highly populated, with the outlying ahupua'a less so.

Sharing skills and knowledge were of pivotal importance to well-functioning ahupua'a. Effective response to periodic droughts, blight, flooding, fire, shifts in the distribution and/or abundance of marine resources, and other challenges required specialized knowledge and adaptive strategies to ensure resource availability during these and other challenges.

Given the variety of ecological attributes, products, and exigencies of life in any given moku or ahupua'a, traditional ecological knowledge and expertise were highly valued social attributes. In the hierarchical sociopolitical system of old Hawai'i, these attributes were expressed and given power in the konohiki. The konohiki both advised and gave deference to higher authority such as the chief of the moku or the island mōʻī (monarch). Notably, intensification of inter-island warfare in the seventeenth and eighteenth centuries (Cordy 2000) may be seen as an important social-organizational force that solidified allegiance to one's island and its leaders.

Konohiki was a special position of responsibility afforded to highly respected individuals (Pukui and Elbert 1986; Malo 1951: 142). It was the konohiki especially who were known to hold specialized and collective ecological knowledge and who made decisions about the proper manner of pursuing, using, and managing resources in the respective ecological zones from mountain to sea, and the specific kapu used to enable or constrain such activities. As described by Andrade (2008), the konohiki played a pivotal role in ancient Hawaiian society, as a repository of ecological knowledge and an intermediary between the people of the land and their leaders:

> In pre-European times, all segments of island society were entirely interdependent. Sources for livelihood and the resources available for use by the people were all contained within the environs of the islands. By today's standards, these resources were very basic, but they were sufficient for the ancestors to have a full and meaningful life as long as they fulfilled their mutual responsibilities to each other and the ʻāina. Konohiki therefore had to possess a wide array of skills. They had to know all of the waiwai (assets) contained within each ahupuaʻa – hydrologic, biologic, and geologic. They had to know the state of the soil, plants, and animals on land and sea, and guide decisions on their use. Most important, konohiki had to know how to deal with human beings. In traditional society, konohiki were bridges connecting the governing and the governed. Konohiki had to gather in the fruits of the ahupuaʻa aliʻi, mōʻī, and akua [god]. However, they needed to ensure that the producers of these fruits, the makaʻāinana were well cared for and fairly treated. If not, according to traditional customs and practices, makaʻāinana were free to move and invest their time and energy under more deserving konohiki, an option easily made possible by the extensive

kinship networks enjoyed by most families extending far outside single ahupua'a...In addition, konohiki had to have respect from the people and enough charisma to draw in and make maka'āinana feel confidence about investing their lives and energy in the long-term success of the ahupua'a. (pp. 75–76)

Foods from the sea were highly valued in nutritional terms and constituted valuable items for trade within and between ahupua'a. As such, konohiki typically possessed extensive knowledge of marine resources and means for harvesting them. Such resources were also given as 'auhau (tribute) to the ali'i (Sahlins 1992: 28). Fish and fishing were central to the functioning of society in old Hawai'i. Various kapu on fishing and use of seafood were established and enforced at the discretion of the konohiki. Such kapu were a particularly compelling form of resource management, since violators could lose various privileges, even of life (Valerio 1985: 231).

Marine resources associated with ko'a (coral) and other bathymetric features were pursued and used as food for residents of the proximate ahupua'a. In some areas, use rights and kuleana (responsibilities) also extended to fishing locations and resources in the offshore waters (Kamakau 1992: 177–178). Even very specific grounds and resources in distant waters of the deep sea could be located by triangulating between landmarks (Kaha'ulelio 2006: 42–61). The words of Kamakau, written in 1867, lend a basic sense of the societal importance of seafood, the cooperative nature of fishing, and the base of knowledge underlying fishing gear and its proper use in old Hawai'i (Kamakau 1992):

Fishing was one of the chief occupations in old days. The fishhooks were made of turtle shell, dog, fish or human bones, prongs of hard wood, and other materials. Fish were caught in deep-sea fishing grounds of a depth of from thirty to forty fathoms, or sometimes of four hundred fathoms... Fishermen went in search of such fishing grounds and learned to locate a particular spot and to return to it again and again. They kept its location a secret from others; it was like a food dish to them...(Fishing) required experts who knew where the schools of fish generally ran...The expert fishermen are most of them dead, and their art is becoming lost to this generation. (p. 239)

Sahlins (1992: 19) asserts that, in the context of the overall ahupua'a, the sea and its resources were particularly important. The products of the uplands were essential for life on the island, but the sea was the point of departure for interisland travel, the place of fish and fishing and, as stated by the author, the realm of the ali'i:

> ...the ahupua'a was able to synthesize the master antithesis of kai and uka, sea and mountain. These categories were ubiquitous in Hawaiian thought and practice, as significant in ritual as in the organization of production. The opposition was asymmetric, however: sea and seaward were privileged in most contexts by relation to the interior of the land. Economically more valued, the coastal areas were also generally preferred for chiefly residence. Here were the most extensive wet taro lands, off-shore and onshore fish ponds, as well as access to the sea and the fishing and surfing that in Hawaii were sports of kings. (ibid.: 19)

Much knowledge of the natural resources of the land and sea was accrued and communicated across generation in old Hawai'i. Such knowledge, precious in its capacity to ensure survival and an ultimate flourishing, was expressed in many and various ways, stories, legends, and everyday dialogue. An attempt to relate the Hawaiian's intimate understanding of the Island world, in this case regarding the continually changing weather conditions that affected fishing and farming across the islands is offered by Handy et al. in their classic text *Native Planters in Old Hawaii* (1972):

> In formulating summarily the science or knowledge of the actual facts that served the [Hawaiian] planter, weather wisdom must be recognized as an element of major importance. Anyone who has known native fish-ermen and their intimate understanding of the signs in the heavens, current, winds, and sea, and native planters and their equally intimate perception of weather signs in clouds, mists, wind, and general atmos-pheric conditions, realized how weather-wise these people are. To others who have not known fishers and planters it is impossible to convey even a hint of the quality of mind and sensory perception that characterizes the human being whose perpetual rapport with nature from infancy has been unbroken. The sky, sea, and earth, and all in and on them, are alive with meaning indelibly impressed upon every fiber of the unconscious and conscious psyche. (p. 23)

Knowledge of land, sea, and sky; oversight of ahupua'a by konohiki and various specialists; and the labor of the maka'āinana allowed for effective management and use of natural resources and their efficient use across Hawai'i of old. Similar systems were developed over time in other parts of the Pacific islands. But as noted by Gonschor and Beamer (2014: 55), "comparative studies of Polynesian land tenure and social organization clearly show that the various islands had developed in different directions of socio-political evolution." The authors provide useful insight into the origins of the Hawai'i approach vis-à-vis those of island societies elsewhere in Polynesia (Gonschor and Beamer 2014):

> Two Polynesian societies, Hawaii and Tonga, had evolved from tribal, kin-based systems of social organization and land tenure to centralized states with feudal-like systems of land tenure, while Tahiti was somewhere in the middle of an evolution between the two. Hawaiian ahupua'a were thus similar geographically to land units in the traditionally organized Polynesian islands but had a different sociopolitical function. While the land units in other islands were socio-familial, they were that *and* territorial in Hawaii. A linguistic analysis of the word ahupua'a—a Hawaiian innovation not cognate to any land division term in other parts of Polynesia, shows its purpose as unit for offering tribute to a centralized government, referring to an [altar]...upon which tribute for the island's ruler and ho'okupu [offerings to make lands productive] would be deposited during the annual makahiki ceremonies. Unlike chiefs heading clans and their territories in [other] traditional Polynesian societies, Hawaiian ahupua'a were administered by konohiki, resource managers appointed by the ruler of large districts of entire islands... According to Hawaiian traditions, the system of ahupua'a divisions was created by rulers who unified or centralized governance of their respective islands, such as Ma'ilikūkahi on O'ahu and 'Umi on Hawai'i Island. (p. 55)

The ahupua'a system of land tenure enabled sustainable use and management of natural resources in the Hawaiian Islands for many centuries. As Hawaiian society was increasingly disrupted through interaction with European explorers and missionaries beginning in the late eighteenth century, so also were the social processes that sustained highly productive use of land and sea, such as construction and maintenance of sophisticated loʻi (terraced irrigation systems) and the expert crafting of fishing hooks, line, and nets.

As elaborated toward the end of the following chapter, by the late nineteenth century, Hawaiians were fishing and cultivating the land primarily for purposes of consumption by the ʻohana, or as a means for earning money in the context of an increasingly dominant cash economy. Ancient strategies for effectively managing and using resources of land and sea persisted in certain forms and places through the Plantation era (Maly and Maly 2003 II) and some continue to be implemented around the Hawaiian Islands today. This persistence of culture into the recent past and present is a central theme of this book, well-expressed by Davianna ʻMcGregor (2007), who describes the people of the land in whom the virtues of the ahupuaʻa continue to be expressed (Fig. 1.4):

Fig. 1.4 Contemporary view of Kualoa, an important ahupuaʻa in Koʻolauloa Moku, Oʻahu

In the context of the native Hawaiian cultural renaissance of the late twentieth century, the word kuaʻaina [person from the back country] gained a new and fascinating significance. A kuaʻaina came to be looked upon as someone who embodied the backbone of the land. Indeed, kuaʻaina are the Native Hawaiians who remained in the rural communities of the islands, who took care of the kūpuna or elders, continued to speak Hawaiian, bent their backs and worked and sweated in the taro patches and sweet potato fields, and held that which is precious and sacred in the culture in their care. The kuaʻaina are those who withdrew from the mainstream of economic, political, and social change in the Islands. They did not enjoy modern amenities and lived a very simple life. This moʻolelo recounts how the lifeways of the kuaʻaina enable the Native Hawaiian people to endure as a unique, distinct, dignified people even after over a century of Americans control of the islands. (p. 4)

References

Anderson, A. J. (1979). Prehistoric Exploitation of Marine Resources at Black Rocks Point, Palliser Bay. In B. F. Leach & H. M. Leach (Eds.), *Prehistoric Man in Palliser Bay* (pp. 49–65). *Bulletin of the National Museum of New Zealand*, Vol. 21.

Anderson, J. L. (2016). An Isolated Tribe Emerges from the Rain Forest—In Peru, an unsolved killing has brought the Mashco Piro into contact with the outside world. *New Yorker*. A Reporter at Large. August 8 and 15, 2016 Issue.

Andrade, C. (2008). *Hāʻena: Through the Eyes of the Ancestors*. A Latitude 20 Book. Honolulu: University of Hawaii Press.

Apple, R. A., & Kikuchi, W. K. (1975). *Ancient Hawaii Shore Zone Fishponds: An Evaluation of Survivors for Historical Preservation*. Office of the State Director. National Park Service. United States Department of the Interior, Honolulu.

Barrere, D. (1961). *Summary of Hawaiian History and Culture*. Hawaiian and Pacific Collection. Hamilton Library. University of Hawaii at Manoa, Honolulu.

Beaglehole, J. C. (Ed.). (1967). *The Voyage of the Resolution and Discovery, 1776–1780*. Cambridge: Hakluyt Society.

Beckley, E. M. (1883). *Hawaiian Fisheries and Methods of Fishing with an Account of the Fishing Implements Used by the Natives of the Hawaiian Islands*. Honolulu: Advertiser Steam Print.

Beckwith, M. (1970). *Hawaiian Mythology*. Honolulu: University of Hawaii Press.

Berkes, F. (1999). *Sacred Ecology: Traditional Ecological Knowledge and Resource Management*. London: Taylor & Francis Group.

Brenzinger, M., & Heinrich, P. (2013). The Return of Hawaiian: Language Networks of the Revival Movement. *Current Issues in Language Planning, 14*(2), 300–316.

Brown, C., Mokuau, N., & Braun, K. L. (2009, July). Adversity and Resiliency in the Lives of Native Hawaiian Elders. *Social Work, 54*(3), 253–261. Special Issue on Practice Perspectives with Racial and Ethnic Minorities.

Bushnell, O. A. (1993). *The Gifts of Civilization: Germs and Genocide in Hawaii*. Honolulu: University of Hawaii Press.

Cachola-Abad, C. K. (1993). Evaluation of the Orthodox Dual Settlement Model for the Hawaiian Islands: An Analysis of Artefact Distribution and Hawaiian Oral Traditions. In M. W. Greaves & R. C. Green (Eds.), *The Evolution and Organizational Prehistoric Society in Polynesia* (pp. 13–32). Auckland: New Zealand Archaeological Association.

Cachola-Abad, C. K. (2000). *The Evolution of Hawaiian Socio-Political Complexity: An Analysis of Hawaiian Oral Traditions*. Dissertation submitted in partial fulfillment of the requirements for the doctoral degree in anthropology at the University of Hawaiʻi at Mānoa, Honolulu. Available at http://www.anthropology.hawaii.edu/people/alumni/pdfs/2000-abad.pdf.

Caviedes, C. (2001). *El Niño in History: Storming Through the Ages*. Gainesville: University of Florida Press.

Clark, G., & Anderson, A. (2014). The Pattern of Lapita Settlement in Fiji. In White, J. P. & P. Sheppard (Eds.), *Archaeology in Oceania*. Sydney: Oceania Publications.

Collerson, K. D., & Weisler, M. I. (2007, September 28). Stone Adze Compositions and the Extent of Anricent Polynesian Voyaging. *Science, 317* (5846), 1907–1911.

Conover, D. O., & Munch, S. B. (2002). Sustaining Fisheries Yields Over Evolutionary Time Scales. *Science, 297,* 94–96.

Cordy, R. H. (1981). *A Study of Prehistoric Social Change: The Development of Complex Societies in the Hawaiian Islands*. Studies in Archaeology. New York: Academic Press.

Cordy, R. H. (2000). *Exalted Sits the Chief.* Honolulu: Mutual Publishing.

Diamond, J. (2005). *Collapse: How Societies Choose to Fail or Succeed.* New York: Viking Press.

Dickinson, W. R. (2009, March). Pacific Atoll Living: How Long Already and Until When? *GSA Today.* Geological Society of America.

Dye, T. (1994). Population Trends in Hawaii Before 1778. *The Hawaiian Journal of History, 28,* 1–20.

Earle, T. K. (1978). *Economic and Social Organization of a Complex Chiefdom: the Halalea District, Kauai, Hawaii.* Ann Arbor: University of Michigan.

Earle, T. K. (1997). *How Chiefs Came to Power: The Political Economy in Prehistory.* Stanford: Stanford University Press.

Erlandson, J. M., Graham, M. H., Bourque, B. J., Corbett, D., Estes, J. A., & Steneck, R. S. (2007). The Kelp Highway Hypothesis: Marine Ecology, the Coastal Migration Theory, and the Peopling of the Americas. *The Journal of Island and Coastal Archaeology, 2*(2), 161–174.

Finney, B. (1994). *Voyage of Rediscovery: A Cultural Odyssey Through Polynesia.* Berkeley: University of California Press.

Finney, B. (2006). Ocean Sailing Canoes. In K. R. Howe (Ed.), *Vaka Moana—Voyages of the Ancestors.* Honolulu: University of Hawaii Press.

Finney, B., & Low, S. (2006). Navigation. In K. R. Howe (Ed.), *Vaka Moana: Voyages of the Ancestors.* Honolulu: University of Hawaii Press.

Firth, R. (1936). *We the Tikopia.* London: George, Allen and Unwin.

Firth, R. (1939). *Primitive Polynesian Economy.* London: George, Allen and Unwin.

Firth, R. (1967). *The Work of the Gods in Tikopia* (2nd ed.). New York: Humanities Press.

Fontaine, M. A. K. (2012). *Two Views of Ancient Hawaiian Society.* A Thesis Submitted to the Graduate Division of the University of Hawaiʻi at Manoa in partial fulfillment of the requirements for the degree of Master of Arts in History, Honolulu.

Fornander, A. (1878). *An Account of the Polynesian Race, Its Origins and Migrations.* London: Tribner and Company.

Glazier, E. W. (Ed.). (2011). *Ecosystem Based Fisheries Management in the Western Pacific.* Hoboken, NJ: Wiley-Blackwell Publishers. ISBN 978-0-8138-2154-2.

Gonschor, L., & Beamer, K. (2014). Toward an Inventory of ahupuaʻa in the Hawaiian Kingdom: A Survey of 19th and Early 20th Century Cartographic and Archival Records of the Island of Hawaii. *The Hawaiian Journal of History, 48,* 53–87.

Goto, A. (1986). *Prehistoric Ecology and Economy of Fishing in Hawaii. An Ethnoarchaeological Approach.* Dissertation submitted in partial fulfillment of the requirements for the doctoral degree in anthropology. University of Hawai'i at Manoa.

Hamilton, C. (2017). *Defiant Earth: The Fate of Humans in the Anthropocene.* Cambridge, UK: Polity Press.

Handy, E. S. C., Handy, E. G., & Pukui, M. K.. (1972). *Native Planters in Old Hawaii—Their Life, Lore, and Environment.* Bernice P. Bishop Museum Bulletin 233. Honolulu: Bishop Museum Press.

Hau'ofa, E. (1994). Our Sea of Islands. *The Contemporary Pacific,* 6(1), 147–161. First published in *A New Oceania: Rediscovering Our Sea of Islands* (V. Naidu, E. Waddell, & E. Hau'ofa, Eds.). Suva, Fiji: School of Social and Economic Development, The University of the South Pacific.

Hau'ofa, E. (2005). The Ocean in Us. In A. Cooper (Ed.), *Culture and Sustainable Development in the Pacific.* Canberra: Australian National University Press.

Henry, T. (1928). Ancient Tahiti. *Bishop Museum Bulletin, 48,* 119–128. Honolulu: Bishop Museum Press.

Holmes, T. (1981). Provisions for Polynesian Voyages. In *The Hawaiian Canoe* (1st ed.). Hanalei, Kaua'i, Hawai'i: Editions Unlimited Publishers.

Hommon, R. J. (1976). *The Formation of Primitive States in Pre-contact Hawaii.* Dissertation submitted in partial fulfillment of the requirements for the Doctorate in Anthropology at the University of Arizona, Tucson.

Hommon, R. J. (2013). *The Ancient Hawaiian State: Origins of a Political Society.* Oxford, UK: Oxford University Press.

Howe, K. R. (Ed.). (2006). *Vaka Moana: Voyages of the Ancestors.* Honolulu: University of Hawaii Press.

Hunt, T. (2007). Rethinking Easter Island's Ecological Catastrophe. *Journal of Archaeological Science, 34,* 485–502.

Hutchings, J. A., & Fraser, D. J. (2007). The Nature of Fisheries- and Farming-Induced Evolution. *Molecular Ecology, 17,* 294–313.

Irwin, G. (1992). *The Prehistoric Exploration and Colonisation of the Pacific.* Cambridge: Cambridge University Press.

Irwin, G. (2006). Voyaging and Settlement. In K. R. Howe (Ed.), *Vaka Moana: Voyages of the Ancestors.* Honolulu: University of Hawaii Press.

Jokiel, P. L., Rodgers, K. S., Walsh, W. J., Polhemus, D. A., & Wilhelm, T. A. (2011). Marine Resource Management in the Hawaiian Archipelago: The Traditional Hawaiian System in Relation to the Western Approach. *Journal of Marine Biology,* 2011, 1–16.

Kahā'ulelio, D. (2006). *Ka oihana Lawai'a: Hawaiian Fishing Traditions.* Originally published in 1902 in *Nupepa Kuokoa* (M. K. Pūkui, Trans. and M. Puakea Nogelmeier, Ed.). Honolulu: Bishop Museum Press and Awaiaulu Press.

Kalakaua, D. (1990). *The Legends and Myths of Hawaii: The Fables and Folklore of a Strange People* (R. M. Daggett, Ed.). Honolulu: Mutual Publishers.

Kamakau, S. M. (1976). *Na Hana a ka Po'e Kahiko* (The Works of the People of Old). Translated from the Newspaper Ke Au 'Oko'a by M. K. Pukui. Arranged and edited by D. Barrere. Bernice Bishop Museum Special Publication 61. Honolulu: Bishop Museum Press.

Kamakau, S. M. (1992). *Ruling Chiefs* (Rev. ed.). Original edition compiled in 1961. Honolulu: Kamehameha Schools.

Kame'eleihiwa, L. (1992). *Native Land and Foreign Desires: How Shall We Live in Harmony?* Ko Hawaii aina a me na koi muumake a ka poe haole: pehea la e pono ai? Honolulu: Bishop Museum Press.

Kamehameha Schools. (1994). *Life in Early Hawai'i: The Ahupua'a* (3rd ed.). Honolulu: Kamehameha Schools Press.

Kana'iaupuni, S. M., Ledward, B., & Malone, N. (2017). Mohala I ka wai: Cultural Advantage as a Framework for Indigenous Culture-Based Education and Student Outcomes. *American Education Research Journal, 54*(1), 311S–339S.

Kana'iaupuni, S. M., Malone, N. J., & Ishibashi, K. (2005, September). *Income and Poverty Among Native Hawaiians—Summary of* Ka Huaka'i *Findings* (PASE Report, 05–06: 5). Kamehameha Schools, Honolulu.

Kane, H. K. (1997). *Ancient Hawaii.* Captain Cook, Hawaii: Hawaii Kawainui Press.

Kawaharada, D. (1999). *Notes on the Discovery and Settlement of Polynesia.* Polynesia Voyaging Society, Honolulu. Available at http://www2.hawaii.edu/~dennisk/voyaging_chiefs/discovery.html.

Kawaharada, D. (2004). *Local Geography.* Honolulu: Kalamakū Press.

Keesing, R. (2005). Do Native Peoples Today Invent Their Traditions? In K. M. Endicott & R. L. Welsch (Eds.), *Taking Sides: Clashing Views on Controversial Issues in Anthropology* (3rd ed.). Dubuque, IA: McGraw-Hill/Dushkin.

Kirch, P. V. (1982). The Impact of the Prehistoric Polynesians on the Hawaiian Ecosystem. *Pacific Science, 36*(1), 1–14.

Kirch, P. V. (1985). *Feathered Gods and Fishhooks: An Introduction to Hawaiian Archaeology and Prehistory.* Honolulu: University of Hawaii Press.

Kirch, P. V. (1997a). Microcosmic Histories: Island Perspectives on "Global" Change. *American Anthropologist, 99*(1), 30–42.

Kirch, P. V. (1997b). Changing Landscapes and Sociopolitical Evolution in Mangaia, Central Polynesia. In P. V. Kirch & T. L. Hunt (Eds.), *Historical Ecology in the Pacific Islands: Prehistoric Environmental and Landscape Change* (pp. 147–165). New Haven: Yale University Press.

Kirch, P. V. (2000). *On the Road of the Winds.* Oakland: University of California Press.

Kirch, P. V. (2010). *When Did the Polynesians Settle Hawai'i?—A Review of 150 Years of Scholarly Inquiry and a Tentative Answer.* Article Based on the Keynote Address Delivered to the Society for Hawaiian Archaeology at the 2010 Annual Meeting at Wailua, Kaua'i. Available at https://www.researchgate.net/publication/260248796_When_Did_the_Polynesians_Settle_Hawai%27i_A_Review_of_150_Years_of_Scholarly_Inquiry_and_a_Tentative_Answer.

Kirch, P. V., & Hunt, T. L. (Eds.). (1997). *Historical Ecology in the Pacific Islands: Prehistoric Environmental and Landscape Change.* New Haven, CT: Yale University Press.

Kirch, P. V., & Sahlins, M. (Eds.). (1992). *Anahulu: The Anthropology of History in the Kingdom of Hawai'i* (Vol. 1). Chicago: Historical Ethnography. University of Chicago Press.

Klein, R. G. (2009). *The Human Career: Human Biological and Cultural Origins* (3rd ed.). Chicago: University of Chicago Press.

Lepofsky, D., & Caldwell, M. (2013). Indigenous Marine Resource Management on the Northwest Coast of North America. *Ecological Processes, 2*(12). A Springer Open Journal.

Levinson, M., Ward, R., & Webb, J. (1973). *The Settlement of Polynesia: A Computer Simulation.* Minneapolis: University of Minnesota Press.

Lyons, C. J. (1875, July 2). Land Matters in Hawaii. *The Islander, 1*(18), 104.

Lyons, C. J., & Alexander, W. D. (1893). The Song of Kuali'i. *Journal of the Polynesian Society, 2,* 161–166.

Malo, D. (1903). *Hawaiian Antiquities (Mo'olelo Hawai'i).* Honolulu: Hawaiian Gazette Company.

Malo, D. (1951). *Hawaiian Antiquities (Mo'olelo Hawai'i)* (2nd ed., N. B. Emerson, Trans.). Bernice P. Bishop Museum Special Publication 2. Honolulu: Bishop Museum Press (Original work published 1898).

Maly, K., & Maly, O. (2003). *Ka Hana Lawai'a A Me Na Ko'a O Na Kai 'Ewalu: A History of Fishing Practices and Marine Fisheries of the Hawaiian Islands* (Vols. I and II). Hilo: Kumu Pono Associates. Prepared for the Nature Conservancy and Kamehameha Schools.

Mandryk, C. A. S., Josenhans, H., Fedje, D. W., & Mathewes, R. W. (2001). Late Quaternary Paleoenvironments of Northwestern North America: Implications for Inland Versus Coastal Migration Routes. *Quaternary Science Reviews, 20,* 301–314.

McGregor, D. P. (2007). *Na Kua 'Aina: Living Hawaiian Culture.* Honolulu: University of Hawaii Press.

Neich, R. (2006). Voyaging After the Exploration Period. In *Vaka Moana, Voyages of the Ancestors: The Discovery and Settlement of the Pacific.* Honolulu: University of Hawaii Press.

NeSmith, R. K. (2005). Tutu's Hawaiian and the Emergence of a New Hawaiian Language. *Ōiwi Journal 3: A Native Hawaiian Journal.* Honolulu: Ōiwi Press.

Nunn, P. D. (2000). Environmental Catastrophe in the Pacific Islands Around A.D. 1300. *Geoarchaeology, 15,* 715–740.

Office of Hawaiian Affairs, State of Hawai'i. (2014). *Income Inequality and Native Hawaiian Communities in the Wake of the Great Recession: 2005 to 2013.* Ho'okahua Waiwai (Economic Self-Sufficiency) Fact Sheet, Vol. 2014, No. 2. Honolulu.

Peiser, B. (2005). From Genocide to Ecocide: The Rape of Rapa Nui. *Energy & Environment, 16*(3, 4), 513539.

Poepoe, K., Bartram, P., & Friedlander, A. (2003). The Use of Traditional Knowledge in the Contemporary Management of a Hawaiian Community's Marine Resources. In N. Haggan, C. Brignall, & L. Wood (Eds.), *Putting Fisher's Knowledge to Work. Fisheries Centre Research Reports, 11*(1), 328.

Polynesia Voyaging Society. (2018). Available at http://www.hokulea.com/.

Pukui, M. K., & Elbert, S. H. (1986). *Hawaiian Dictionary* (6th ed.). Honolulu: University of Hawaii Press.

Rainbird, P. (2002). A Message for Our Future? The Rapa Nui (Easter Island) Ecodisaster and Pacific Island Environments. *World Archaeology, 33,* 436–451.

Rollett, B. V. (2002). Voyaging and Interaction in Ancient East Polynesia. *Asian Perspectives, 41*(2), 182–194.

Sahlins, M. D. (1958). *Social Stratification in Polynesia.* Seattle: University of Washington Press.

Sahlins, M. D. (1992). *Anahulu: The Anthropology of History in the Kingdom of Hawai'i,* Volume 1, *Historical Ethnography* (P. V. Kirch & M. Sahlins, Eds.). Chicago: University of Chicago Press.

Sibert, J., & Hampton, J. (2003). Mobility of Tropical Tunas and the Implications for Fisheries Management. *Marine Policy, 27,* 87–95.

Stannard, D. E. (1989). *Before the Horror: The Population of Hawai'i on the Eve of Western Contact.* Honolulu: University of Hawaii Press.

Swain, D. P., Sinclair, A. F., & Hanson, J. M. (2007). Evolutionary Response to Size-Selective Mortality in an Exploited Fish Population. *Proceedings of the Royal Society B., 274,* 1015–1022.

Taonui, R. (2006). Polynesian Oral Traditions. In K. R. Howe (Ed.), *Vaka Moana: Voyages of the Ancestors.* Honolulu: University of Hawaii Press.

Trask, H. (2005). Natives and Anthropologists: The Colonial Struggle. In K. M. Endicott & R. L. Welsch (Eds.), *Taking Sides: Clashing Views on Controversial Issues in Anthropology* (3rd ed.). Dubuque, IA: McGraw-Hill/Dushkin.

Tuggle, H. D., Cordy, R., & Child, M. (1978). Volcanic Glass Hydration-Ring Age Determination for Bellows Dune, Hawaii. *New Zealand Archaeological Association Newsletter, 21,* 57–77.

United States Government Accountability Office. (2014). *American Samoa and the Commonwealth of the Northern Marianas Islands.* Economic Indicators Since Minimum Wage Began. Report to Congressional Committeee. GAO-14-381. Washington, D.C.

Valerio, V. (1985). *Kingship and Sacrifice: Ritual and Society in Ancient Hawaii* (P. Wissing, Trans.). Chicago and London: The University of Chicago Press.

Vitousek, P. (1995). Why Focus on Islands? In P. Vitousek, L. L. Loope, & H. Andersen (Eds.), *Islands: Biological Diversity and Ecosystem Function.* Ecological Studies 115. New York: Springer.

Waihona 'Aina Corporation. (2006). *Boundary Commission Data.* Kailua, Hawai'i.

Young, K. G. T. (2012). *Rethinking the Native Hawaiian Past* (2nd ed.). Series on Native Americans: Interdisciplinary Perspectives. New York and London: Routledge, Taylor & Francis Group.

2

Sociocultural Change and Persistence During the Historic Period

2.1 Overview

The course of history that preceded contemporary Native Hawaiian interests in using and managing natural resources of island and ocean is particularly lengthy, and it necessitated adaptation to profound social and environmental challenges along the way. As discussed here, socioeconomic challenges became acute after contact with foreigners who sought to alter indigenous ways of life and profit from the commercial value of the islands and their natural resources.

All Polynesian societies were forced to address social change upon contact with foreign explorers, missionaries, traders, and entrepreneurs. Most early interactions were less than ideal, partly because ethnocentrism and a false air of moral and racial superiority prevented the visitors from fully appreciating and respecting the ancient and unique island societies and cultures they were encountering. Moreover, the foreigners brought and parlayed material items and concepts that were often as damaging as they were helpful.

© The Author(s) 2019
E. W. Glazier, *Tradition-Based Natural Resource Management*,
Palgrave Studies in Natural Resource Management,
https://doi.org/10.1007/978-3-030-14842-3_2

In New Zealand, initial interactions between explorers and natives were problematic and often violent from the start. As in Hawai'i, many Māori turned to Christianity when epidemics of introduced disease ravaged the islands. But many indigenous residents subsequently lost respect for the missionaries when they acquired and sold land to a growing number of pākehā (white settlers). The impacts of, and resistance to colonization of Aotearoa were substantial from the start and continue in the twenty-first century (Middleton 2015; Tuhiwai-Smith 1999; Duff 1990).

In Hawai'i, the Kānaka Maoli (the original colonists of Hawai'i and their descendants) tended to adopt that which was likely to suit their needs and interests and to resist that which was oppressive. As examples, iron was readily adopted for many practical uses, new weaponry was used for interisland warfare, and sandalwood was traded by ali'i (chiefs) for cash, goods, and services. When the missionaries took on the task of converting spoken Hawaiian to a written language, the Hawaiians used printing presses to document mo'olelo (stories) and na'auao (wisdom) from generations past, thereby preserving indigenous knowledge and tradition.

But Hawaiian authors of the nineteenth century also recognized and related the many profound problems that accompanied haole as they visited and eventually sought to recolonize the islands. As noted by Silva (2004: 21), Kamakau's original mo'olelo of Cook's arrival in Hawai'i described "recurrent violence, mainly on the part of Cook's men," and that "the ali'i's kahuna advised against retaliating … emphasizing that welcoming the foreigners … was the pono thing to do." The Hawaiian perspective on subsequent problems introduced by haole visitors to Hawaii was succinctly expressed in metaphorical terms at the concluding section of Kamakau's account of Cook's visit to Hawai'i, written in the early 1860s, as provided in *Ke Kumu Aupuni*:

'O nā hua a me nā 'ano'ano o kāna mau hana I kanu ai, 'ōmamaka, mai nō ia a ulu, a lilo mau kumulau hoolaha e ho'olaha e ho'oneo ai i ka lāhui kanaka o kēia mokupuni.

[The fruits and the seeds that his [Cook's ocis] actions planted sprouted and grew and became trees that spread to devastate the people of these islands]. (Kamakau 1996: 45)

Kamakau (1996: 45) goes on to list the introduced "fruits and seeds" that had become so damaging to his people: (1) gonorrhea together with syphilis; (2) prostitution; (3) the false idea that he [Cook] was a god and [to be] worshipped; (4) fleas and mosquitoes; (5) the spread of epidemic diseases; (6) change in the air we breathe; (7) weakening of our bodies; (8) changes in plant life; (9) changes in the religions, put together with pagan religions; (10) changes in medical practice; and (11) laws in government.

Of all such changes, introduced disease was most profound and hardest to address. As discussed further along in this chapter, diseases for which the long-isolated islanders had little immunity took a terrible toll in indigenous communities across Polynesia, and many islanders subsequently questioned or rejected missionary teachings that so clearly promised a benevolent god and eternal life.

Significantly, Hawaiians have practiced their own powerful forms of spirituality and have maintained particularly rich oral traditions throughout the course of time. As such, Hawaiians' own beliefs and lessons from the past, coupled with exogenous concepts or information they deemed useful or meaningful, helped guide them through the changes and tribulations of the eighteenth, nineteenth, and twentieth centuries.

The deep Polynesian past and its lessons were never abandoned by the Kānaka Maoli. Even when the survival of an ancient society was in question during the nineteenth century, Hawaiian culture was sustained by the maka'āinana (commoners) and ali'i and their dedication to the words and wisdom of the kūpuna (ancestors) whose knowledge and work led them to Hawai'i and enabled the development of a flourishing society. As indicated by native authors such as Andrade (2008) and McGregor (2007), and as discussed in this and the following chapters, central aspects of Hawaiian culture were internally strengthened rather than rendered obsolete following contact with European and American explorers, missionaries, and entrepreneurs.

2.2   First Encounters and Misconceptions

The first and early arrival of foreign explorers and missionaries in Hawai'i and elsewhere in Polynesia led to fundamental problems among indigenous island societies. In most Polynesian archipelagos, cultural incompatibilities were associated with immediate or eventual resistance to foreign visitors. There was often much cultural misinterpretation on both sides at first contact.

For instance, Māori resistance to Dutch incursions in Aotearoa during the mid-seventeenth century was immediate and forceful (Horwitz 2002: 102–104) and it preceded centuries of troubled cultural interactions with "colonizing" Europeans (Tuhiwai-Smith 1999). Resistance was similarly immediate and persistent in Samoa when and after the French first set foot there in 1787 (La Pérouse 1995). Likewise, the arrival of Wallis in Tahiti in 1767 quickly turned violent, "with a number of cultural misunderstandings between the Tahitians and the British from the very start" (Claessen 1997: 185). Pearson (1969) describes both how the visitors failed to understand the Tahitians' welcoming rituals, and the reactions of the islanders to mistakes in cultural protocol made by the visitors.

Notably, Hawaiians were initially accepting of newly arriving visitors—with some scholars asserting that the explorers were first seen, or at least treated as, "sacred chiefs and envoys of the divine world" (Tcherkezoff 2008: 3; see also Sahlins 1995; Borofsky 1997). The exploring parties, meanwhile, tended to be less open-minded, if not condescending and suspicious of the island people they were encountering. As noted above, violence was initiated by the explorers during the period early contact. In Hawai'i, an archipelago that was (and is) of ongoing interest to foreign powers, indigenous resistance was expressed quickly and over the course of time in ways both overt and subtle. Activist resistance to what are seen as ongoing incursions against a sovereign island nation continues among many Kanaka Maoli Native Hawaiians in the present day.

Tcherkezoff (2008) describes the cultural basis of problems between Europeans and Polynesians during the era of exploration. Unfortunately, such contexts, differences, and resulting difficulties were characteristic of indigenous-exogenous interactions of the era throughout much of Oceania. In essence, neither the foreign visitors nor the indigenous islanders seemed aware or appreciative of cultural relativism and its value as a concept for forging peaceful and mutually beneficial relations between societies that were dissimilar in fundamental but not irreconcilable ways. As Tcherkezoff explains (2008):

> …preconceived views blinded [European explorers] …and prevented them from understanding the whole range of acts and behaviors of the indigenous population. In particular, several contexts were the results of gross misunderstandings… [For instance]…A complete misunderstanding about rules of nakedness, where European taboos of the time were projected indiscriminately onto the inhabitants, together with the close association between nakedness and sexuality that prevailed in European ideology, meant that the indigenous population was generally qualified as "lascivious, lewd" and the female seen as engaging in wanton behavior…European visitors were unaware that, on the contrary, such behaviors took their meaning as part of very formal dances with the fact of presenting oneself in the finale without clothing as a mark of "respect" towards [the] visitors. (pp. 2–3)

Various misunderstandings so induced were to have lingering effects for explorers and islanders across Oceania. Unfortunately, what might have become lasting mutually beneficial relationships between dissimilar societies quickly became fraught with tension, unrest, and effects that worsened as ever greater numbers of sailing ships appeared on the horizons of the Pacific islands.

2.3   Introduced Items and Their Effects

Given that the population of Hawai'i was burgeoning under a system of sociopolitical relations that, at point of first encounter with Europeans may well have developed into a full nation-state (Hommon 2013), subsequent decades of inordinate difficulties, including those ultimately challenging the very survival of the Hawaiian people, originated not internally, but rather in what the foreigners brought to the islands, and the manner of their bringing. Knowingly and unknowingly, the haoles brought concepts and items that were indeed foreign. Some were adopted for beneficial use by the islanders, others proved detrimental; some highly so.

After Cook's arrival on Kaua'i in 1778, and his death on the shores of Kealakekua Bay on Hawai'i Island in 1779, explorers came in succession—Portlock, Dixon, Vancouver, LaPerouse, Lisiansky, and others. These were followed by fur traders and whaling vessels and their crews, sandalwood buyers, other entrepreneurs, and subsequently the missionaries. This is discussed by Archer (2018):

> After Cook was killed, the ports went quiet. For seven years no one came. Then, in 1786, two separate fleets arrived within days of each other. Never again would a year pass without visitors from abroad (Kuykendall 1938:20). Shortly, the islands became the principal stopover for merchants in a new Pacific fur trade that Cook himself had anticipated. Hailing from Britain, France, Spain, and by 1800 overwhelmingly the United States, Pacific traders pulled into any safe Hawaiian harbor that would accept them. Most harbors, most of the time, accepted them. Popular ports such as Kealakekua Bay on the Big Island, Waimea Bay on Kaua'i, and La Pérouse Bay on Maui quickly developed into international transit points with Hawaii and its people caught up in global networks of exchange. (p. 54)

As reported by Greene et al. (1993: III-C-3), certain material items brought to the islands during the exploratory and entrepreneurial forays of the eighteenth and nineteenth centuries were of value to the Hawaiians, who readily accepted them for practical use. Iron, for instance, was quickly recognized for its practical value, and its acquisition by Hawaiians appears to have been one of the factors leading to the shoreside skirmish in which Captain Cook lost his life (Stokes 1931: 6).

Jarves (1843b: 112) notes that by "1790, firearms, gunpowder, and liquor had become prized trade items in the islands," and that "articles of warfare were in demand, and abundantly supplied by thoughtless traders, who in some cases found them turned upon themselves." Notably, arms and gunpowder acquired from trading vessels were used with great effect to conduct warfare, especially by Kamehameha I, who after years of war, negotiation, and adept political maneuvering, finally gained ascendance over ruling chiefs across the islands. Kamehameha ultimately established united monarchical rule in 1810, albeit with underlying dissension until and following his death in 1819.

It is noteworthy that, despite his witnessing the demise of Cook at the hands of Hawaiians as a junior officer on Cook's fateful third visit to Hawai'i, Captain George Vancouver later established positive relationships with certain ali'i [including Kamehameha I] and expressed pleasure in returning to Hawai'i after ventures elsewhere in the Pacific. According to Archer (1987: 52), Vancouver wrote that Hawai'i was "an asylum, where the hospitable reception and friendly treatment were such as could not have been surpassed by the most enlightened nation of the earth" (Fig. 2.1).

Positive political interactions aside, the island landscape was dramatically altered after Vancouver released livestock on the main islands in the early 1790s. Notably, Kamehameha placed a ten-year kapu on consumption of ungulates so that the creatures could multiply for use in the years to come (Speakman and Hackler 1989: 57). As discussed by Silva (2004), who cites Kuykendall (1938), Vancouver was well-intentioned if ecologically ignorant in this action, believing "that the flora and fauna of England were superior [to those of Hawai'i], and left cattle and several kinds of fruits and vegetable plants to improve the landscape" (pp. 23–24) "Unfortunately," Silva notes (ibid., p. 24), "these acts only served to upset the ecological balance".

Injury to Hawaiian society and the island landscape imparted by explorers and later arrivals may not have been intentional, but neither was the potential for long-term damage sufficiently contemplated. Again, an air of superiority on the part of the visitors led to fundamental problems. Although the Hawaiians themselves altered ecosystems across the islands during original colonization and during periods of population growth and expansion, it should be kept in mind that the changes brought by the newly arriving visitors impacted a well-established system of production in the ahupua'a, with detrimental implications for a sovereign people that continue to play out in the twenty-first century. Problems resulting from the introduction of ungulates to the islands, and Hawai'i especially, are profound and persistent, as described by Chynoweth et al. (2010):

Fig. 2.1   A sketch of Kealakekua Bay in 1864 by Missionary Rufus Anderson

Beginning in the late eighteenth century, Europeans brought a variety of species to the Hawaiian Islands, many of which have subsequently established feral populations. The original purpose of some vertebrate introductions was likely to populate oceanic islands with a food source to access during later voyages (Campbell and Donlans 2005). Other animals became established after arriving as stowaways on ships, or more recently as result of the purposeful introduction of game animals. Domestic goats (*Capra hircus*) were introduced to provide food for sailors on long voyages, but they quickly became a self-sufficient feral population (Yokum 1967). Non-native feral goats have a tremendous impact on island ecosystems.

The Hawaiian Islands are particularly vulnerable to the detrimental impacts of non-native ungulates, due to their high degree of geographic and evolutionary isolation and large proportion of endemic species. Overall, goats are considered to be "the single most destructive herbivore introduction to island ecosystems globally". (King 1985) (p. 1)

Hawaiians made practical use of many introduced items and became adept traders in their own right, having in their possession much practical knowledge, great agricultural and fishing skills, and associated commodities in hand. As discussed by Greene et al. (1993) and supporting literature, certain items were highly valued by the indigenous residents:

> During the early Western contact period, Hawaiian farmers were able to increase the production of goods and commodities to meet the traders' demands and satisfy the needs of the *ali'i* without a major dislocation of island economics. Hawaiians quickly learned the value of their goods and showed a strong ability to barter. On Hawai'i [Island], early traders found plentiful sugarcane, breadfruit, coconut, plantain, sweet potatoes, taro, yams, bananas, and hogs, as well as introduced oranges, watermelon, muskmelon, pumpkin, cabbages, and garden vegetables (Townsend 1888)…It was not long before Hawaiians began to demand clothing, cloth, pitch, flour, and other western products. As described by one trader, "the islanders…ceased to care for objects of mere ornament, and preferred in their traffic cloth, hardware and useful articles". (Jarves 1843b: 125) (p. III-C-3)

Of note, although the introduction of distilled alcohol is said to have been problematic for Hawaiians (Owen 1898)—as it is for all human societies—the substance was distributed with apparent abandon by shrewd and knowing American and European traders (Kelly 1993: 371). Hopkins (1866), for instance, states that the early traders "commenced implanting among the chiefs the taste for ardent spirits" (p. 149).

Alcohol is said to have had significant implications for Hawaiian society after the death of Kamehameha I, when his son and successor, Liholiho, apparently influenced by distilled spirits, yielded to intrafamilial political pressure to proclaim the 'ai noa, or end to the long-held 'ai kapu (prohibitions on certain foods and manner of eating), which at the time dictated that only men could cook, that men and women consume food separately, and that certain foods be prohibited for consumption by women. This along with the failure of Kekuaokalani to restore the 'ai kapu by warfare were seen by many Hawaiians as significant blows to the "sanctity of tradition" (Silva 2004: 27–29; Kamakau 1996: 27–45) (Fig. 2.2).

Fig. 2.2   Ancient kiʻi pōhaku (petroglyphs), Moku o ʻEwa, Oʻahu

Although European explorers observed and documented the physical grandeur of Polynesia, they clearly failed to grasp the grandeur of Polynesian *societies* and what had been accomplished in the settlement of these most distant archipelagos. This was the combined effect of imperialist designs on foreign lands and resources and an egregious form of ethnocentrism on the part of haole leaders. As discussed by Joseph (2016: 587), lack of understanding or lack of will to understand or appreciate complex island societies and the cultural attributes of indigenous islanders might be expected, given the ultimate intent of the early explorers to lay claim to new lands, and the norm of societal superiority that characterized the culture and demeanor of those representing expanding empires of the era (Joseph 2016):

> Beginning around the eighteenth century, the desire to expand across the globe and acquire territories was prolific for the leading world powers. As large empires began to explore uncharted territories, a simultaneous desire to stake a claim in those territories also occurred. Prestige for the individuals who discovered new lands was eminent; however, the empire or country that financed these explorations had an economic stake in what was discovered as well. In the process of exploration, many first explorers of new lands encountered native peoples with a different way of life and different culture. This is the point where ethnocentrism makes an appearance and shapes what happens next. (p. 587)

2.4   Missionary Enterprise and Ethnocentrism

In true ethnocentric fashion, haole (*papālagi* in Samoan, *popa'a* in Tahitian, and *pakeha* in Māori) explorers, whalers, and traders neglected to consider or address the social and health implications of their own presence and actions in the islands. This certainly also held true for the missionaries, who arrived in Polynesia not long after reports of inhabited Pacific islands reached America and Europe. The missionaries exhibited a particularly self-righteous form of cultural and religious certainty, sailing to Polynesia with directed purpose to alter the islanders and their ways of life, and in the case of Hawai'i, ultimately using Hawaiian labor for material gain.

Ethnocentrism is well-exemplified in the psychological hegemony that characterized the strategies of the earliest missionaries serving in the islands during the 1820s. Theirs was an unapologetic mission to "save" the Hawaiians from the "darkness [that] covered the earth and [the] gross darkness of the people" (Bingham 1855: 2).

While the Hawaiians were prepared to see the newly arriving haoles in a positive light (Sahlins 1995), the missionaries came to the islands already "knowing." They came with the intent of "correcting," of rendering an ancient society into something like their own. Little did the missionaries perceive or choose to accept and respect:

a. the trials, skills, failures, successes, and persistence of the many hundreds of generations that preceded and enabled first arrival of Polynesians in this most distant island chain many centuries prior;
b. the intrepid nature of, and myriad challenges encountered by Hawaiians as they navigated to, discovered, and settled the islands; their development of a highly efficient system of use of land and sea; and their ultimate forging of a system of sociopolitical relations as sophisticated as any of that human era;
c. the complex nature of the Hawaiian cosmology; the holistic nature of the Hawaiian system of religious beliefs; and the spiritual linkages between the Hawaiian, the land, the sea, the gods, and the ancestors; or
d. the innate capacity of indigenous Hawaiians to equal or exceed persons of any other ancestral background on earth in any measure of intellectual, physical, spiritual, or other achievement.

In short, the missionaries were ignorant of the great society they were hoping to mold into something else. As indicated in the memoirs of Hiram Bingham I (1855), leader of the first American Protestant missionaries to introduce Christianity to Hawai'i, misperception of Polynesians and Polynesian history, coupled with a sense of moral dominance, hegemony, and ethnocentrism characterized the disposition of New England missionaries toward their Polynesian "subjects":

> In the place of being filled with love and reverence to the true God, and equity and benevolence towards his creatures, they are "filled with all unrighteousness, fornication, wickedness, murder, debate, deceit, malignity," being "whisperers, backbiters, haters of God, despiteful, proud, boasters, inventors of evil things, disobedient to parents, without natural affection, implacable, unmerciful." Such was the character of the famed "children of nature" or "children of wrath of nature"

at the Sandwich, Society, New Zealand, and Marquesas Islands, while they had not been taught by inspired truth, to stand in all of the holiness, power and justice of the Maker, Law-giver, Redeemer, and Judge of the world. The process by which children, born of heathen parents, come to possess a character so odious, and so fearfully at variance with the laws of their Moral Governor, and with the design of man's creation, deserves our attention and care, especially if it be possible for us to arrest it. In the peculiarities of national character and condition, of the Hawaiians and other heathen tribes, ought to be studied and delineated in the process of evangelizing the world; in order to show the adaptation, and make the successful application, of the Gospel to the wants of idolaters, wherever they dwell. (p. 23)

Significantly, and handily, the code of morality espoused by the missions to Polynesia also held that humans and human societies must be industrious in the strictest sense of the word, i.e., that labor must be of the Calvinist sort—without interruption, ordained by God, and contributing to the advancement of civilization. As discussed by Weber (1930), such "advancement" contributes to, and is defined in part by, capitalist enterprise. This is particularly significant in the Hawai'i case, since the dictum of individual labor yielded major benefits for the missionaries, many of whom used the toil of the maka'āinana to become wealthy agricultural barons after private ownership of land was enabled by the Alien Land Ownership Act of 1850. Such was one outcome of the arrival of Christian missionaries and their doctrines in early nineteenth century Hawai'i—with effects that linger into the present day, as certain descendant missionary families continue to own large parcels of land and control capital that was first generated through the toil of Hawaiians long ago.

2.5 Early Cultural Resistance

The introduction of new concepts and materials both benefited and challenged an ancient island society. But as made so clear by scholars such as Osorio (2002) and Silva (2004), Hawaiians skillfully resisted and/or engaged new items and ideas to suit their needs (and continue to do so, as discussed in the following chapters). Hawaiians were, for example, initially circumspect about the religious tenets being purveyed by the missionaries—as would be expected of a people practicing their own highly developed system of religious beliefs and living as part of a spiritual order that is said to interpenetrate virtually all aspects of life—one in which the human and divine are deeply entangled and the very 'āina (land), underlying kūpapakū (bedrock), moana (ocean), and auoli (sky) are imbued with mana (divine power) and cultural meaning.

Silva (2004: 33) uses the Hawaiian concept of pono (balance, completeness) to illustrate alterations of meaning that occurred when the missionaries converted Hawaiian into a written language, so that the tenets of Christianity could be communicated to the islanders. The oral-to-written language task was not easy, partly because the missionaries found Hawaiian difficult to understand and speak (Kashay 2002), and partly because certain Hawaiian concepts and meanings were not readily translated into the English language. Silva (2004) reveals that although Hawaiians gradually adopted a written form of their own language and many took on Christian beliefs, neither the correct meanings of spoken Hawaiian nor underlying indigenous beliefs or spiritual practices were abandoned:

> Whereas pono had been used previously to describe the ideal behavior of ali'i and other concepts such as balance, completeness, and material well-being, it now took on the foreign connotation of conforming to Christian morality. Rightness meant something different in the two worlds, but the missionaries were able, in barely questioned translation, to appropriate this powerful term—an appropriation that would have radical consequences...But translating the term pono into Christian righteousness was a trying task for the missionaries. Most Kānaka did not enter as quietly into that foreign and restricted lifestyle as some histories would have us believe. Bingham's memoirs are brimming with complaints about the struggle with indolent, pleasure seeking [Hawaiian]. Indeed, his chapter about the second year of the mission starts this way: "While some of the people who sat in darkness were beginning to turn their eyes to the light, others with greater enthusiasm were wasting their time in learning practicing or witnessing the *hula*." (Bingham 1855: 123 as cited in Silva 2004: 33)

Ironically, Bingham and other missionaries failed to grasp or respect the fact that hula (dance) was (and remains) a core spiritual practice of indigenous Hawaiians and that perception of licentious movement (or perception that sexuality itself could not be experienced as sacred) lay in the eye of the beholder—in this case, in the eyes of puritanical evangelists. Such perceptions, admonishment of indigenous practitioners, and alteration or reduction of the meaning of spoken words (such as pono) to fit Christian ideals, in fact, did very little to diminish the true and actual meaning of an ancient language or the underlying cultural significance of such deeply held ways of life and expressions of spiritual relationships between Hawaiians, their gods, the land, and surrounding ocean. As indicated by Silva (2004: 12), and by Chun (2006), "pono" encapsulates numerous Hawaiian values and actions that are not readily reduced to a single meaning (see also Chun's translation of Malo in Malo and Chun 1996).

The concept of pono and the practice of hula remain central aspects of Native Hawaiian culture to this day. While public performance of hula was verbally banned by Queen Ka'ahuman in 1830, it continued to be performed and practiced until restrictions were eased in 1870, and ultimately publicly embraced by King Kalakaua in 1883 (Chronicling America 2018). The real origins and meaning of hula clearly are spiritual in nature, as expressed, for example, by Mahealani Uchiyama in her recent book *The Haumana Hula Handbook for Students of Hawaiian Dance* (2016):

> …Hōpoe learned to sway her hips by imitating the movement and rhythm of the surf … From this time forward, the hula was offered to the gods, done in honor of Hawaii's royalty, and danced in celebration of life's mysteries. Thus, we understand that the hula was received by the people as a result of closely observing their natural environment: the play of the wind through the trees, the sound of the ocean as it breaks on the shore—all are understood to be sacred gifts imparting comfort, wisdom, and inspiration. These natural forces were internalized and manifested in the bodies of the ancestors and the result was the hula. (pp. 4–5)

The missionaries may also have neglected to comprehend that, in developing and proffering the Hawaiian language in written form to facilitate "salvation" (Bingham 1835), they also advanced a mechanism that Native Hawaiians could use to enhance communication between themselves. For instance, detailed letters could be delivered relatively quickly to one or more people on other islands without the speakers (writers) ever leaving the point of origin. Significantly, the language gradually served as a tool for social coalescence and collective resistance to colonial powers. The significance of this phenomenon is discussed by Silva (2004), who asserts that:

> The print media in the vernacular contributed to the imagination of the nation among people who did not know each other personally, but now shared a large community (Anderson 1991). The lāhui [nation of indigenous Hawaiians] was also created in the collective imagination by Kanaka Maoli grouping themselves as alike, sharing a language and culture, albeit with regional variations, and in opposition to the haole. That opposition was not simply an othering based on differences in color and language, but an attempt to fend off U.S. and various European colonial advances. Hawai'i, the nation of the lāhui did not exist as a singular entity before the arrival of foreigners; it was, rather, Hawai'i, Maui, O'ahu, and so on. Moreover, newspapers and literacy introduced the Kanaka Maoli to anti-colonial struggles around the world. (p. 88)

As further discussed by Silva (2004), the subtleties of spoken Hawaiian were not lost when the language was converted for use in textual form, and communication via the new written language could be as subtle and resistant to haole objectives as the Kanaka Maoli wished or needed to make it:

> ...the Kanaka Maoli, both men and women, maka'āinana and ali'i, speak in a variety of ways, some of which were not understood and not *meant to be* understood by the colonizers. The word "kaona"...[refers to] hidden meaning, as in Hawaiian poetry; concealed reference, as to a thing, person, or place; [and] words with double meanings (Pukui and Elbert 1986:130). It is a well-known characteristic of the Hawaiian language that is generally spoken of in reference to mele but that also is common in writing and in everyday speech. An awareness of the political function of kaona, especially the possibilities for veiled communication,

helps in analyzing the words and actions of the Kanaka Maoli ... [Kaona] was crucial in creating and maintaining national solidarity against the colonial maneuvers of the U.S. missionaries, the oligarchy, and the U.S. politicians. Without knowledge of the cultural codes in Hawaiian, foreigners who understood the language could still be counted on to miss the kaona. (p. 8)

Hawaiian in written form thus provided not only a venue for serving the evangelistic intent of the haole missionaries, but also an important mechanism for indigenous resistance to imposed change. While the psychological hegemony of the missionaries and the introduction of products by traders and entrepreneurs were disruptive of traditional lifeways and landscapes in certain ways and places, Native Hawaiians were resilient, akamai (smart), and deeply attached to the ways of the past and their own unique culture in the present. Heartfelt or immediate conversion to introduced religion was by no means universal. Further, as noted by Silva (2004: 85) "many [Hawaiians] did not experience the conflict between the ancient beliefs and their Christianity that the missionaries expected or wanted them to experience ... [and, for instance] some reconciled conflicts by comparing people and events in the ancient tradition with the ones described in the Bible."

Thousands of years of evolving indigenous religious traditions are not easily or quickly abandoned, no matter the message or means of evangelism. As such, the concept of syncretism (blending of religious tenets) may be a more accurate term regarding shifting beliefs among most Hawaiians of the period. Some Hawaiians did convert to Christianity, in part due to fears that disease and mass death were a form of punishment for moral transgression—as instilled so condescendingly by certain missionaries (Herman 2010: 271). On the other hand, scourges of epidemic disease ultimately led many Hawaiian to question or abandon introduced forms of religion altogether. The dire nature of the Mormon mission during the epidemics of the mid-nineteenth century provides a good example of the situation, as captured in a diary entry by elder Thomas Karren (as cited in Kenney 1997):

A cloud of gloom [is] hanging over this place. Those crowded and spirited meetings which were carried on hear a few months ago, has [sic] disappeared. Our meetings in a great measure has [sic] been broken up. We have had to give up our meeting house. So great has been the distruction [sic] among this People that the[y] all most dispare [sic] of life. Even them that Servive [sic]. (p. 13)

Kenney (1997) goes on to report that both Protestant and Mormon missions made economic demands on their converts, and that this also led to widespread disenchantment:

The summer of 1853 was faith shattering. Those who had joined the church believing the priesthood would protect them and their loved ones were devastated ... But the epidemic was only one cause for the decline of Mormonism on Oʻahu. Many early converts had joined because the Mormons imposed no financial burden on members, as opposed to the Protestants, who required substantial donations to support missionaries and their families, build and maintain schools and churches, and even sponsor other missions in the Pacific. When the Mormons introduced tithing in October 1852, many [Hawaiians] were disillusioned. (p. 14)

As Silva (2004: 13) argues so adeptly, resistance to cultural changes such as those proffered or foisted by the missionaries was given a venue for expression when Hawaiians developed and operated their own newspaper (*Ka Hoku ka Pakipika*, established in 1861). Hawaiian culture and tradition were also communicated when moʻolelo and subtle messages were made available in newspapers supported by the missions, such as the long-lasting *Nupepa Kuakoa* (Silva 2004):

First, the Kanaka Maoli of the nineteenth century, in addition to strenuously resisting every encroachment on their land and lifeways, consciously and continuously organized and directed their energies to preserving the independence of their country. Second, throughout the nineteenth century, print media, particularly newspapers, functioned as sites for broad social communication, political organizing, and the perpetuation of the native language and culture. It has often been noted that the change from orality to literacy has eroded native forms of thought and expression, especially due to the fixing and consequent reduction in possible meanings and versions of text. Hereniko has written, for example, that "the written word has undermined the fluidity of indigenous history. Oratory allowed for debate and negotiation ... [while] the written word fixes the truth" (Hereniko 2000:82). For the Kanaka 'Ōiwi [native sons] of Hawai'i nei [this Hawai'i], however, who observed the passing away of their relatives and friends and genocidal numbers, writing, especially newspapers, was a way of ensuring that their knowledge was passed on to future generations Silva. (p. 13)

2.6   Introduced Disease

Hawaiians successfully resisted, negotiated, and/or engaged a wide variety of externally imposed sources of change in the decades following the arrival of Cook on the shores of Kaua'i in 1778. But most unfortunately, newly introduced diseases were particularly difficult to withstand, especially given (a) the lethality of first epidemics among Polynesian populations long isolated from the global pathogen pool (Shanks 2016); and (b) the limited state of understanding of bacterial and viral infections and their effective treatment prior to the twentieth century (penicillin was first developed in 1928).

As discussed by Shanks (2016), retrospective epidemiological analysis strongly suggests that bacillary dysentery, probably associated with *Shigella* spp., was particularly deadly for Hawaiians and other Polynesians during the late eighteenth and early nineteenth centuries. Dysentery, which involves an inflammatory immune response, tissue damage in the colon, impaired absorption of nutrients, and excessive dehydration, among other symptoms, underlay massive population loss in Hawai'i, Tahiti, Fiji, Samoa, and other archipelagos across the Pacific. Notably, dysentery followed first exposure to certain viral infections among Pacific islanders, such as measles, thereby dramatically increasing rates of mortality from such forms of disease.

Though other scholars suggest more conservative figures, the work of Stannard (1989) indicates that some 800,000 or more Native Hawaiians may have been living across the archipelago when Cook arrived. But as the author notes (Stannard 2000) "whatever the population [size] at the moment of Western contact, it did not remain that large for long," and "in their lengthy isolation, the Hawaiians had been spared the ravages of every major epidemic disease (smallpox, typhoid, yellow fever, measles, and others) that had long infected much of the rest of the world, and they had no previous exposure to treponemic infections (such as syphilis) or to tuberculosis." As such, first interactions with European explorers initiated a long and tragic period of disease and population loss among an ancient and theretofore isolated society (Stannard 2000):

> Venereal disease and tuberculosis were already spreading among the Hawaiians by the time Cook's ships departed the islands for good in 1779. Seven years later the surgeon on board a French frigate that visited a remote part of Maui wrote that the majority of natives there had become infected with venereal disease. And six years later still, Captain George Vancouver…revisited the islands and reported finding massive depopulation virtually everywhere (Rollin 1799, Volume I:341 and Volume II, pp. 337-338; Vancouver 1798, Volume I:158-160, 187-188). From this point on, account after account by European and American explorers reported on the visible ravages of newly-introduced disease among the Hawaiians, and the precipitous decline of this formerly robust population. (p. 1)

Indicative of the scope of deadly effects resulting from newly introduced disease, a dysenteric epidemic called ʻōkuʻu (crouching disease) is said to have killed many thousands of persons on Oʻahu in 1804, as Kamehameha I readied to invade Maui (see Archer 2018). Stannard (1989: 55) also refers to this disease (maʻi ʻōkuʻu), stating that it may have reduced the population of Native Hawaiians living on Oʻahu prior to the 1823 census by as much as 75%. Amundson and Ruddle-Miyamoto (2010) use population figures to indicate the catastrophic loss of Hawaiian lives during the subsequent decades:

> In 1832 the native population was about 130,000. In 1850 it was about 84,000. In 1900 there were about 28,000 native Hawaiians and 8,000 part-Hawaiians [remaining in the islands]. (Section 4)

It is important to note that, although disease originated in various harbors and population centers frequented by foreigners, few island areas were spared from epidemics since rural residents of the early nineteenth century increasingly travelled to Honolulu, Lahaina, Kailua, and other towns for purposes of trade, acquisition of goods and services, and eventually for employment. Moreover, as throughout the course of Hawaiian history, maka'āinana traveled across ahupua'a, moku, and on occasion even to other mokupuni (islands) to visit relatives, friends, or members of other 'ohana. As such, the epidemics were fueled in part by social interaction. As reiterated by Jarves (1843a: 17) "it occasioned a great mortality—so much so that the natives state [that] the living were not able to bury the dead."

The immediate impacts of severe illness among Native Hawaiians in the eighteenth and nineteenth centuries are difficult to envision. But as discussed by Stannard (2000), the demographic implications of ongoing population loss and the constraining effects of certain disease on fertility impacted the indigenous population in ways both overwhelming and persistent:

> Ghastly as these scourges were, the most devastating long-term damage done to the islands' population occurred in the wake of these epidemics and, additionally, as a consequence of widespread tuberculosis and venereal disease among the people. Year after year, with virtually no periods of relief, disease-caused infertility and subfecundity produced recorded birth rates so low that, even in the absence of epidemics, missionaries and others routinely predicted the imminent extinction of the Hawaiians. (p. 1)

Schmitt and Nordyke (2001) note that neither the adept kahuna lapa'au (Hawaiian medical specialist) nor physicians from afar were equipped to effectively remedy the epidemics and their debilitating symptoms among members of afflicted 'ohana across the islands:

> Much of the burden of care fell on the kāhuna la'au lapa'au, the native practitioners. Although viewed with disdain by the foreign doctors, these practitioners were not noticeably inferior to their foreign colleagues in treating most disorders and accidents. However, the kahuna [priest, expert in medicine] lacked experience with such diseases as now threatened the Islanders. (p. 6)

The epidemics initiated by foreign visitors to Hawai'i fundamentally reduced the capacity of Hawaiian 'ohana to operate ahupua'a as they had in decades and centuries past. This is not to say that the land was no longer worked, that marine resources were no longer harvested by the Hawaiians, or that well-tested resource management strategies were not followed. At a time when so many were afflicted and in need, these activities and strategies very likely became *more* important, if more challenging, than ever. Rather, the epidemics caused great stress to the entire population, and the integrated nature of the ahupua'a-based system of resource production was concomitantly stressed.

Periods of epidemic disease, protracted recovery, and long-term demographic effects (Stannard 1990) unavoidably meant that ancient ways of life on the 'āina began to shift. Moreover, massive population loss rendered broadly scaled sociopolitical resistance to the haole and haole-induced changes increasingly difficult (Table 2.1). Basic survival took priority.

Quite obviously, Hawaiians of centuries past experienced great and persistent psychological stress, sadness, and anxiety due to foreign disease, mass death, and other major changes (Salzman 2012: 30). In the case of the stigmatizing haole response to Hansen's disease in the nineteenth and twentieth centuries, forced exile and the racial dynamics of colonialism can be seen as constituting a unique violation of human rights (Amundsen and Ruddle-Miyamoto 2010).

The profound impacts of disease and its introduction by foreigners whose objectives had little or nothing to do with the well-being of the original colonists of Hawai'i were expressed by Kamakau in a passage that was ultimately redacted from *Ruling Chiefs* (Silva 2004: 26). Kamakau had, over time, become disillusioned with haole ways and concepts, and "rather than interpreting the mass deaths as the results of God's anger at the lack of piety of the Kanaka" [as the missionaries and those unduly influenced by the missionaries would have the people believe], Kamakau directly blamed foreign ways (Silva 2004):

Table 2.1 Major outbreaks of disease in the Hawaiian Islands: 1778–1853[a]

Dates	Types of disease	Affected islands	Notes
1778–1779	Syphilis, gonorrhea	All	Associated with reduced fertility
1804	ʻŌkuʻu	Oʻahu	Dysentery probably associated with typhoid fever or cholera
1818	Unidentified respiratory	Oʻahu	–
1824–1825	Unidentified respiratory	Oʻahu	Possibly influenza, high rate of mortality
1826	Whooping cough	Maui, Oʻahu	–
1834	Unidentified respiratory	Maui, Oʻahu	–
1844–1845	Influenza	Oʻahu	–
1848–1849	Measles, influenza, whooping cough	Hawaiʻi Island, Kauaʻi, Maui, Niʻihau, Oʻahu	As many as 10,000 dead
1853	Smallpox	Hawaiʻi Island, Kauaʻi, Maui, Oʻahu	As many as 6000 dead, led to disillusion with missionaries (Kenney 1997: 13)
Late 1800s	Hansen's disease (leprosy)	Various islands, patients forcibly moved to Kalaupapa, a remote peninsula on Molokaʻi	First diagnosed in Hawaiʻi in 1848

[a]After Archer (2018: 59)

O ke kumu i loaʻa mai ai kēia pōʻino a me ka hoʻoneo ʻana hoʻi i ka lāhui Hawaiʻi nei, ua maopopo, ʻo nā haole nō ka poʻe pepehi lāhui; a ʻo ka puni hanohano me ka puni waiwai, ʻo lāua nō nā hoa aloha no ka maʻi luku.

[The reason for the misfortune and the decimation of the Hawaiian lāhui (nation), it is understood, is that the haole are people who kill other peoples; and the desire for glory and riches, those are the companions of the devastating diseases]. (p. 26)

2.7   The Shift to Private Ownership of Land and Its Effects on the Makaʻāinana

During the period of population expansion and rise of complex socio-political relations in old Hawaiʻi, life in the ahupuaʻa was one of production and reciprocal transfer of goods, services, protection, and leadership in a hierarchy of makaʻāinana, konohiki, kahuna (priest, expert), and aliʻi. There was also sufficient time and numerous venues for familial and communal interaction, religious and spiritual activities, and various forms of recreation, such as surfing the majestic swells of Hawaiʻi, of which much is written (see Clark 2011).

According to Silva (2004: 40), "land tenure was the central feature of this system of social and political relationships, based on obligations as well as bonds of affection." The authority of the konohiki or aliʻi to govern and receive ʻauhau (tribute) required that he care for the resources of ocean and land in a wise and equitable manner, establishing kapu and kānāwai (rules) to maintain balance between use and conservation of resources as necessary for near- and long-term survival.

Subsequent to the reign of Kamehameha, and that of his son Liholiho (who reigned from 1819 to 1823), the old ways of social organization and governance began to falter. As Sahlins (1992: 3) notes, no sooner had the political and economic unity needed for centralized kingdom been achieved than a de facto oligarchy of powerful aliʻi landholders begin to encroach on "royal powers, royal lands, and the king's control of commercial trade"—largely commercial trade of sandalwood to American entrepreneurs who shipped the timber to China. Unfortunately, the sandalwood resource was rapidly exhausted at the expense of maka'āinana labor, a situation that diminished the political legitimacy of the chiefs involved.

Kamehameha's son, Kauikeaouli (Kamehameha III), reigned from 1824 to 1854, a time during which Protestant missionaries advanced a constitutional monarchy that was readily influenced by foreign interests. Despite the king's best attempts to do what he deemed best for the Hawaiian kingdom vis-à-vis its increasing exposure to world powers and economies, the period was characterized by strategic maneuvering on the part of haole advisors and investors and, ultimately, diminishing indigenous control of the kingdom (Sahlins 1992: 3). The political machinations underlying this shift undoubtedly were furtive, occurring in the upper echelons of power during a period of ongoing indigenous struggles with epidemic disease which, as noted above, made overt and widespread resistance on the part of the polity unlikely at best. According to Sahlins (1992):

> White men (haole) took it [the Hawaiian kingdom] over and turned the government into a constitutional monarchy. Progressively undermining the ruling chiefs, the haole officers of the kingdom also attempted to reconfigure the kingship into a subtropical caricature of European royalty. It was as if they wished to make the king doubly symbolic, as much a clown prince as he was a mere figurehead. By the late 1840s the central government was for all intents in the hands of whites, mostly Americans. By the end of the 1850s, the rural areas likewise were largely under white control. In both cases, American Protestant missionaries took the lead, a change of ministries as it were, but they were eventually superseded by haole bureaucrats of kinds and interests—the interests notably of American capital. The mahele or land reform of [the mid-1850s] gave private and inalienable rights to Hawaiians, while at the same time complementary laws were allowing "aliens" to gain titles in fee simple. The combination proved fatal to native Hawaiian ownership of the soil and the integrity of native Hawaiian society. (p. 3)

The mahele (formal apportionment of land), to which Sahlins refers, culminated in 1848, followed by the Kuleana Act of 1850. The great mahele provided that, after the mōʻī (king; Kamehameha III) had retained his private land holdings, the lands of Hawaiʻi would be allotted in thirds, with one-third awarded to the Hawaiian government, one-third to the aliʻi, and one-third to the makaʻāinana. As discussed by Andrade (2008: 80–82), who cites Kuykendall (1938): "by the time the legislation was actually enacted, however, the mōʻī, the government, and the aliʻi (konohiki) possessed the bulk of the land between them." The makaʻāinana ultimately received only 28,000 acres, or much less than one percent of the islands (MacKenzie 1991: 8 as cited in Andrade 2008: 82). Andrade (2008: 82) discusses factors that led to such a small allotment of land to those who best knew, worked, and depended on ahupuaʻa resources around the islands:

> Many makaʻāinana were not aware of the importance of owning their land, having never experienced this unfamiliar practice that contrasted sharply with the lifeways of the past centuries. No one explained to them that, should they fail to obtain their own lands and should the aliʻi or konohiki later lease or sell ahupuaʻa lands on which their homes were located, new owners could evict the makaʻāinana from their homes...The new laws provided only two years for the people to register their claims. One Hawaiian writer of the time said that it would have been much more appropriate if the time was twenty years. Another reason given was the government's failure to inform a population still ill at ease with the written word. Still another factor was the lack of money... Most important, however, was the fear that by taking an award, makaʻāinana might be limiting their own access to the waiwai (resources) of the ahupuaʻa...It is entirely plausible that some makaʻāinana assumed that if they did not act to change anything, the traditional system would simply continue.

Although the mōʻī and aliʻi nui (great aliʻi) are said to have attempted to protect the interests of the makaʻāinana by making government lands affordable and allowing flexible claim deadlines (Silva 2004: 42–43), the factors described above, in conjunction with passage of the Kuleana Act, which enabled foreigners to purchase land in fee simple, put the people of the land in a precarious position.

Andrade, too, asserts that the mōʻī sought to protect the people of the land (2008: 82). But he also notes that the king was erroneous in his assumptions that "these newly enacted laws would protect the makaʻāinana, allowing them to make an adequate livelihood by exercising their traditional use rights within the system of customs that had been practiced on ahupuaʻa lands during the preceding centuries" (Andrade 2008: 82).

Undoubtedly, Kamehameha III could see the world powers of the day were vying for possession of the islands and their strategic and economic value. Seeking to protect the interests of the kingdom, he "took the advice of the new kāhuna (Kameʻeleihiwa 1996: 155) – missionaries and business advisors from England and America, that such a transformation of land management would act to restore the population, would inspire the makaʻāinana to work hard for material rewards in the capitalist economy being put into place, and would ease commercial treaty-making with other members of the family of nations" (Silva 2004: 42).

But as further discussed by Silva (2004), who cites Jonathan Osorio's powerful work *Dismembering Lāhui—A History of the Hawaiian Nation to 1887* (Osorio 2002), the shift to a formal-legal arrangement of land use in the islands was, in part, a matter of co-opting semantics. In the end, the makaʻāinana were left with a future in which their basic rights to extract food from the land and sea as they had for centuries was at best uncertain (Fig. 2.3):

> In the process, makaʻāinana were reconceived as hoaʻāina, which then was translated as "tenants" (Osorio 2002: 52); konohiki became landlords; and traditional obligations and bonds were replaced by laws...Even the word "auau" was co-opted into the new system as "taxes" (ibid., p. 54). While the specific "rights of tenants" were protected under the new law, the makaʻāinana and konohiki became more vulnerable to losing their means of making a living and were only able to exercise gathering rights as specified in the law. (p. 42)

Formal laws favoring haole interests had begun to replace the ancient arrangements that had nurtured Hawaiian civilization for centuries. Like the epidemics—indeed, in combination with disease and the

Fig. 2.3 Kaniakapūpū—ruins of the summer palace of Kamehameha III, Nuʻuanu, Oʻahu (1996)

increasingly dominant cash economy, the mahele and related legalities made life on the land for the makaʻāinana quite different and significantly more challenging than it had been just a half-century before.

Inasmuch as the traditional linkages of mutual responsibility between the people of the land and the aliʻi had now been fundamentally altered, ahupuaʻa could no longer operate in a holistic manner as in the past. Far fewer Hawaiians in total were available to work the land, and new laws and tax burdens displaced many thousands of ʻohana from their homes and ancestral ahupuaʻa. As discussed by Handy and Pukui (1972), many Hawaiians moved from the land to find employment or were otherwise disrupted in their ancient ways of life by in-migrating peoples:

Other influences, in Kaʻū as elsewhere, potent in separating individuals from homeland have been: in the early days, the recruiting of native sailors for whaling; the establishment of the sugar plantations; ranching, which inevitably had the effect of disrupting native agriculture; the taking over of commercial fishing and favorable fishing localities by [other groups]; the substitution of rice for taro in irrigated lands (lo‘i), and of Oriental for Hawaiian labor where taro continued to be grown; and last but by no means least, the steady flow of Hawaiians to Honolulu, able-bodied men seeking work on the waterfront, in the public services and government offices, both sexes being drawn thither for purposes of being taught and of teaching, elders coming to be with their young folks, and many lured from the country by the excitement, sociability and conveniences of city life. (pp. 15–16)

Andrade (2008: 98) too, discusses the departure of maka‘āinana from ancestral lands, in this case as motivated by the need to survive in the new cash economy:

Many maka‘āinana, especially those not granted land, also left their ahupua‘a in post-Mahele times. Often, they settled in towns or near seaports where cash, almost entirely derived from foreign sources, was more available. The need to access sources of cash was a major reason people left the ahupua‘a. They could no longer pay their debts and taxes with products of subsistence life ways or satisfy their obligations to society through contributions of shared labor, as they had in the traditional system. (Andrade 2008: 98 cites Merry 2000: 98–99)

Stover (1997) discusses a variety of additional factors and processes that led to failed claims or people leaving their ‘āina after the mahele. These include but are not limited to: dying without heirs to title; deserting a given claim or claims; failing to cultivate the land, willingly giving up the ‘āina (due to illness or infirmity, for instance); and failing to fulfill pō‘alima obligations (to work on the chief's plantation). Handy and Pukui (1972) express regret about this aspect of Hawaiian history and question how it might have progressed had the haole crafters of the mahele and related laws been attentive to and acted with Hawaiian culture first in mind:

One of the chief motives behind the generous but unwise granting of individual title to lands on the part of the ali‘i in the Great Mahele was the hope of keeping Hawaiians on their land. It may be that one reason why that hope has not been realized was because the titles were granted to individuals. Country Hawaiians in the middle of the nineteenth century had little understanding of the meaning of private ownership. But they had a very strong sense of *family rights and responsibilities*. It is interesting to speculate on what might have been the outcome of the Mahele had the grants been as 'ili or sections allocated to 'ohana, represented legally by their respective haku, instead of as parcels (kuleana) in fee simple to individuals. Probably in most instances the haku would have been guided in decisions, planning and action by the whole 'ohana [and] he would at least have been subject to the advice of the family and of shrewd and hard-headed elders in particular. (p. 17)

2.8   The Beginnings of the Plantation Era

In true capitalist fashion, cheap labor and strategic investment of funds generated during the whaling era and through various enterprises in American and Britain, contributed to the emergence of five large business entities in Hawai'i, most with missionary roots. Sugar was the primary commodity of production, and large tracts of land were acquired to facilitate the overall industry. These firms controlled most commerce around the islands for many decades after their incorporation: C. Brewer & Company, Ltd. (established in 1826); American Factors (1849); Castle & Cooke (1851); Alexander & Baldwin, Ltd. (1860s); and Theo. H. Davies Company, Ltd. (1870). Each of these entities continue to operate in some manner to the present day.

In their description of ancient and changing ways of life in Moku o Ka'ū on Hawai'i Island, Handy and Pukui (1972: 15–16) make clear that indigenous Hawaiians reacted in differing ways to new social and economic situations encountered during and after the mid-nineteenth century. Members of some 'ohana continued to engage a subsistence-oriented way of life on the land, often as tenants who generated cash to supplement the household economy in whatever manner feasible (see Andrade 2008). Others adjusted by leaving the 'āina to seek paying opportunities in Honolulu or other population centers (Handy and Pukui 1972). Yet others worked as laborers on the sugar plantations.

The labor situation in total forced members of many 'ohana to be separated for long periods of time, adding to the range of externally induced challenges.

Indicative of the cultural divide between the Hawaiians and haole entrepreneurs during the initial years of sugar production in Hawai'i, the earliest known labor dispute in the islands involved Native Hawaiian workers at Kōloa sugar mill on the island of Kaua'i striking for higher pay. The year was 1841. The nature of the dispute is described by the National Park Service (2018):

> In 1841, Ladd & Company's Kōloa sugar plantation was the site of the first general strike by native laborers in the Hawaiian Islands. The workers were paid 12.5 cents a day and went on strike to demand an increase in pay to 25 cents per day. The Kōloa plantation management refused, stating that in addition to their base pay the workers were receiving housing, fish, and land for their taro patches as well as an exemption from paying taxes to the *ali'i* (native chiefs). The strike was broken within two weeks. This was the first of many such strikes that affected sugar plantations in the islands throughout the nineteenth and twentieth centuries.

Recruited by haole plantation owners and "managed" primarily by haole luna (foremen), persons of non-Hawaiian ancestry came to the islands in periodic waves of immigration to work the fields during the mid- and late nineteenth century and well into the twentieth century. Ironically, while indigenous Hawaiians cultivated sugar cane centuries before the arrival of European explorers, and haole firms ultimately generated great wealth from the production and distribution of sugar, it was Chinese entrepreneurs who initiated the industry in Hawai'i. According to Nordyke and Lee (1989):

> In 1802, Wong Tze-Chun, a Chinese entrepreneur, brought a mill and boilers to Hawai'i aboard a sandalwood trading ship and established on Lāna'i the first commercial effort toward sugar production (Char 1975: 37). Two men, Ahung and Atai, introduced a sugar mill on Maui in 1828, and other new Chinese residents manufactured sugar on Kaua'i in the mid-1830s and on the island of Hawai'i in the 1840s. By 1838, between 30 and 40 Chinese were counted among the 400 foreigners residing in Honolulu. (Char 1975: 54) (pp. 197–198)

Plantation workers and their families gradually immigrated from Japan, the Philippines, the Azores, and Puerto Rico, among other points of origin. Such persons competed with Hawaiians for plantation work and brought with them new languages, customs, and genetic attributes. Marriage between Native Hawaiian and newly arriving immigrants became increasingly common during the late nineteenth century.

A new form of language developed in the islands during the Plantation era, facilitating communication between the growing number of resident culture groups. Originally termed ʻōlelo paʻiʻai, or more formally Hawaiian-Creole-English, the lexicon has evolved over time into what is commonly called "pidgin." This continues to function as an important form of language in the islands (Higgins 2015; Sato 1993).

The latter part of the nineteenth century in Hawaiʻi can be characterized as a period of rapid economic growth facilitated by foreign acquisition and use of lands for cultivation of sugar. This came at the expense of well-being among many indigenous ʻohana and their capacity to work the land for subsistence and survival. According to Herman (2010), the upsurge of the plantation economy following the mahele changed the overarching intent of the missions, from saving Hawaiians through conversion, to the enabling of economic growth and self-accrual of profit and capital:

> Descendants of missionary families thus easily acquired large tracts of now-available land on which to develop plantations. These changes became much more significant after the signing in 1875 of a Reciprocity Treaty with the United States that allowed Hawaiian sugar to be sold tax-free to the United States. Within a few years, the acreage under sugar [production] doubled, and by 1887 the islands' sales to the United States had jumped from 547 tons to 100,000 tons per year. The political repercussions of this transformation would eventually overthrow the Hawaiian government and produce the unconsciously ironic phrase, "Sugar is king," and [although the population of Native Hawaiians was stabilizing]… discourses of depopulation contributed to that process. (p. 272)

2.9   E Hoʻomau

The great turmoil caused by imperialist exploration, introduced disease, condescending evangelism, and a shift toward capitalist-based trade and labor notwithstanding, the period of radical change following first encounter between haole and Kanaka Maoli in no way spelled the end of traditional knowledge or the attention of Native Hawaiians to the naʻauao of the past, the cultural needs and practices of the day, or consideration of generations to come. Again, a beneficial if unintended consequence of the period was the capacity of Hawaiians to preserve moʻolelo, mele, oli, and ʻike ponolia (practical knowledge) in a written form that could be retained for present and future reference and use.

It would seem that a society so continually challenged by radical change could not readily survive the course of time. But Hawaiians did persist, the population did restabilize by the end of the 1880s (Herman 2010: 273), and core aspects of Hawaiian culture were preserved despite the ravages of the nineteenth century and further macro-social changes in the twentieth (McGregor 2007). Continual waves of immigration, suppression of the Hawaiian language, two world wars, the rise of a massive tourism economy, and new socioeconomic challenges of many sorts did little to disrupt fundamental continuity in key elements of Hawaiian culture.

E. S. Craighill Handy and Mary Kawena Pukui, scholars and students of Hawaiian lifeways in the middle and latter part of the twentieth century, dedicate a section of *The Polynesian Family System in Kaʻu Hawaiʻi* to the issue of "disintegration" (Handy and Pukui 1972). Tragic in its summary discussion of the radical changes that had occurred in the islands over the 150 years prior to the book's writing in the 1950s, the section also indicates clear continuity of Hawaiian cultural practices and a readiness on the part of many Hawaiians to adopt new ways while retaining elements of the old. The authors (ibid.) especially emphasize the ongoing tenacity of the Hawaiian ʻohana—and therein may lie the tenacity of Hawaiian society as a whole:

> With its slow disintegration over a period of 150 years, individuals belonging to dis-articulated ʻohana have had recourse to two means of personal and social adjustment: some, especially those who have clung

to planting and fishing, in other words to the old means of subsistence, and certain families in whom, despite removal from their native milieu, the instinct of loyalty to their Hawaiian progenitors has remained very strong, have clung to their 'ohana and its inclusive obligations and privileges. Others, particularly those most affected by intermarriage and by American education, or by city life, have chosen wholeheartedly to adhere to the American system. Most Hawaiians nowadays adhere to some degree to both the old Hawaiian system and the American. (p. 16)

The authors go on to emphasize the strong and persistent linkages that continued to be observed between the Kanaka Maoli and the Hawaiian Islands, the surrounding sea, and each other. So it was just prior to designation of statehood in 1959:

A great proportion of the Hawaiian people still have their roots in the country, even those who reside in the city; an attachment to the soil and the sea that nourished them is a profound instinct [cultural norm] met with amongst all true Hawaiians and Polynesians everywhere. Equally, attachment to blood relatives is with most Hawaiians a deep and unchanging instinct [cultural norm]. One has to know intimately country Hawaiians, or city Hawaiians who remain Hawaiian at heart, to realize the inner depth and strength, and the outer importance in current living, that blood relationship has today. (Handy and Pukui 1972: 16)

Dark era though it was, and greatly reduced in number, Hawaiians survived untold trials in the centuries following the arrival of Cook and other foreign explorers. Through overt and subtle resistance to unwanted aspects of foreign ways, perseverance in the face of change, and use of that which was new and beneficial, indigenous Hawaiian culture and society would continue to evolve even as the Hawaiian monarchy was threatened by, and eventually illegally overthrown by American forces in 1893, and annexed by the U.S. government in 1898. As discussed in the following pages, these and subsequent imperialist processes were endured by descendants of the original colonists of the Hawaiian Islands, many of whom continue to practice and evolve age-old traditions and customs in this new century (Fig. 2.4).

Fig. 2.4 Hawaiian women working a loʻi on the island of Oʻahu, ca. 1890 (Photograph by Frederick George Eyton Walker 1890)

References

Amundson, R., & Ruddle-Miyamoto, A. O. (2010). A Wholesome Horror: The Stigmas of Leprosy in 19th Century Hawaii. *Disability Studies Quarterly, 30*(3–4).

Anderson, R. A. (1864). *The Hawaiian Islands: Their Progress and Condition Under Missionary Labors.* Boston: Hould and Lincoln.

Anderson, B. (1991). *Imagined Communities: Reflections on the Origin and Spread of Nationalism.* New York: Verso Books Publishing.

Andrade, C. (2008). *Hāʻena: Through the Eyes of the Ancestors.* A Latitude 20 Book. Honolulu: University of Hawaii Press.

Archer, C. I. (1987 Spring). The Voyage of Captain George Vancouver: A Review Article. *BC Studies: The British Columbia Quarterly, 73,* 41–61.

Archer, S. (2018). *Sharks Upon the Land: Colonialism, Indigenous Health, and Culture in Hawaii, 1778–1855.* Cambridge, UK: Cambridge University Press.

Bingham, H. A. M. (1835). *O Ke Kumumua na na Kamalii; He Palapala E Ao aku I na Kamalii Ike Ole I ka Heluhelu Palapala*. Oahu: Mea Pai Palapala Na Na Misionari.

Bingham, H. A. M. (1855). *A Residence of Twenty-One Years in the Sandwich Islands or the Civil, Religious, and Political History of Those Islands: Comprising a Particular View of the Missionary Operations Connected with the Introduction and Progress of Christianity and Civilization Among the Hawaiian People* (3rd ed., Revised and corrected). Canandaigua, NY: H.D. Goodwin, Auctioneer.

Borofsky, R. (1997). Cook, Lono, Obeyesekere, and Sahlins. *Current Anthropology, 38*(2), 255–282.

Campbell, K., & Donlans, C. J. (2005). Feral Goat Eradication on Islands. *Conservation Biology, 19*(5), 1362–1374.

Char, T. (1975). *The Sandalwood Mountains*. Honolulu: University of Hawaii Press.

Chronicling America. (2018). *Chronicling America: Historic Newspapers from Hawaii and the U.S.: Hula*. Honolulu: University of Hawaii at Manoa Library. Available at https://guides.library.manoa.hawaii.edu/c.php?g=105252&p=687126%20.

Chun, M. N. (2006). *Pono: The Way of Living*. Ka Wana Series. Honolulu, HI: Curriculum Research and Development Group.

Chynoweth, M., Litton, C. M., Lepczyk, C. A., & Cordell, S. (2010, February 22–25). Feral Goats in the Hawaiian Islands: Understanding the Behavioral Ecology of Nonnative Ungulates with GPS and Remote Sensing Technology. In R. M. Timm & K. A. Fagerstone (Eds.), *Proceedings of the 24th Vertebrate Pest Conference* (pp. 41–45). Sacramento, CA and Davis, CA: University of California.

Claessen, H. J. M. (1997). The Merry Maidens of Matavai: A Survey of the Views of Eighteenth-Century Participants Observers and Moralists. *Bijdragen tot de Taal-, Land- en Volkenkunde, 153*(2), 183–210.

Clark, J. R. K. (2011). *Hawaiian Surfing: Traditions from the Past*. Honolulu: University of Hawaii Press.

Duff, A. (1990). *Once Were Warriors*. New York: Vintage Press.

Greene, L. W., Rhodes, D. L., & Van Horn, L. F. (1993). *A Cultural History of Three Traditional Hawaiian Sites on the West Coast of Hawai'i Island*. U.S. Department of the Interior, National Park Service. Available at https://www.nps.gov/parkhistory/online_books/kona/history0.htm.

Handy, E. S. C., & Pukui, M. (1972). *The Polynesian Family System in Kau Hawaii*. Rutland, VT and Tokyo, Japan: Charles E. Tuttle Company.

First published in 1958 by the Polyesian Society (Inc.), Wellington, New Zealand.

Hereniko, V. (2000). Indigenous Knowledge and Academic Imperialism. In R. Borofsky (Ed.), *Remembrance of Pacific Pasts: An Invitation to Remake History*. Honolulu: University of Hawaii Press.

Herman, R. D. K. (2010). Out of Sight, Out of Mind, Out of Power: Leprosy, Race, and Colonization in Hawai'i. *Hūlili: Multidisciplinary Research on Hawaiian Well-Being, 6*, 271–301.

Higgins, C. (2015). Earning Capital in Hawai'i's Linguistic Landscape. In R. Tupas (Ed.), *Unequal Englishes*. London: Palgrave Macmillan.

Hommon, R. J. (2013). *The Ancient Hawaiian State: Origins of a Political Society*. Oxford, UK: Oxford University Press.

Hopkins, M. (1866). *Hawaii: The Past Present, and Future of Its Island Kingdom: An Historical Account of the Sandwich Islands, Polynesia* (2nd ed.). London: Longmans, Green.

Horwitz, T. (2002). *Blue Latitudes*. New York, NY: Picador.

Jarves, J. J. (1843a). *History of the Hawaiian or Sandwich Islands*. Boston: Tappan and Dennet.

Jarves, J. J. (1843b, August). The Sandwich or Hawaiian Islands. *The Merchants' Magazine, 9*(2), 112.

Joseph, P. (2016). *The Sage Encyclopedia of War: Social Science Perspectives*. Thousand Oaks, CA: Sage.

Kamakau, S. M. (1996). Ke Kumu Aupuni ka mo'olelo Hawai'i no Kamehameha Ka Na'i Aupuni a me kāna aupuni i ho'okumu ai. In P. Nogelmeier (Ed.), *Ke Kumu Lama, 'Ahahui 'Ōlelo Hawai'i*. Honolulu: Kamehameha Publishing.

Kame'eleihiwa, L. (1996). *He Mo'olelo Ka ao'o Kamapua'a: A legendary Tradition of Kamapua'a: the Hawaiian Pig-God*. Honolulu: Bishop University Press.

Kashay, J. F. (2002). "O That My Mouth Might Be Opened": Missionaries, Gender, and Language in Early 19th Century Hawai'i. *The Hawaiian Journal of History, 36*, 41–58.

Kelly, M. (1993). Testimony Provided to the International Tribunal on the Rights of Indigenous Hawaiians. In W. Churchill & S. H. Venne (Eds.), *Islands in Captivity: The International Tribunal on the Rights of Indigenous Hawaiians* (Vol. II, pp. 370–372). Cambridge, MA: South End Press.

Kenney, S. (1997). Mormons and the Smallpox Epidemic of 1853. *The Hawaiian Journal of History, 31*, 1–26.

King, W. B. (1985). Island Birds: Will the Future Repeat the Past? In P. J. Moors (Ed.), *Conservation of Island Birds: Case Studies for the Management of Threatened Island Species* (pp. 2–3). Cambridge: ICPB Technical Publication 3, International Council for Bird Preservation.

Kuykendall, R. S. (1938). *The Hawaiian Kingdom* (Vol. 1). Honolulu: University of Hawaii Press.

La Pérouse, J-F. G. (1995). *Journal of Jean-François de Galaup de Lapérouse, 1785–1788* (J. Dunmore, Trans. and Ed.). London: Hakluyt Society.

MacKenzie, M. K. (Ed.). (1991). *Native Hawaiian Rights Handbook.* Honolulu: Native Hawaiian Legal Corporation.

Malo, D., & Chun, M. (1996). *Ka Moʻolelo: Hawaiian Traditions.* Honolulu: First People's Press.

McGregor, D. P. (2007). *Na Kua ʻAina: Living Hawaiian Culture.* Honolulu: University of Hawaii Press.

Merry, S. E. (2000). *Colonizing Hawaii: The Cultural Power of Law.* Princeton, NJ: Princeton University Press.

Middleton, A. (2015). Missionization, Maori, and Colonial Warfare in Nineteenth Century New Zealand. In J. Simonds & V. Herva (Eds.), *The Oxford Handbook of Historical Archaeology.* Oxford: Oxford Handbooks Online. Scholarly Research Reviews.

National Park Service. (2018). Old Sugar Mill of Kōloa—Kōloa Hawaiʻi. Component of the *Asian American and Pacific Islander Heritage Discover Our Shared Heritage Travel Itinerary.* Produced by the National Park Service Heritage Education Services, in Partnership with the National Conference of State Historic Preservation Officers. Washington, DC. Available at https://www.nps.gov/nr/travel/asian_american_and_pacific_islander_heritage/Old-Sugar-Mill-of-Koloa.htm.

Nordyke, E., & Lee, R. K. C. (1989). The Chinese in Hawaii: A Historical and Demographic Perspective. *The Hawaiian Journal of History, 23,* 196–216.

Osorio, J. K. (2002). *Dismembering Lāhui: A History of the Hawaiian Nation to 1887.* Honolulu: University of Hawaii Press.

Owen, J. A. (1898). *The Story of Hawaii.* London and New York: Harper & Brothers.

Pearson, W. H. (1969). European Intimidation and the Myth of Tahiti. *The Journal of Pacific History, 4,* 199–217.

Pukui, M. K., & S. H. Elbert. (1986). *Hawaiian Dictionary* (Rev. ed.). Honolulu: University of Hawaii Press.

Rollin, M. (1799). Dissertation on the Inhabitants of Easter Island and the Island of Mowee. In *A Voyage Round the World Performed in the Years 1785, 1786, 1787, and 1788* by the Boussole and Astrolabe under the Command of J. F. G. de la Pérouse. Published by the Order of the National Assembly under the Superintendence of L. A. Milet-Mureau (Translated from the French). London: A. Hamilton.

Sahlins, M. D. (1992). *Anahulu: The Anthropology of History in the Kingdom of Hawai'i. Vol. 1: Historical Ethnography* (P. V. Kirch & M. Sahlins, Eds.). Chicago: University of Chicago Press.

Sahlins, M. D. (1995). *How "Natives" Think: About Captain Cook, for Example.* Chicago: University of Chicago Press.

Salzman, M. (2012). Ethnocultural Conflict and Cooperation in Hawai'i. In D. Landis & R. D. Albert (Eds.), Chapter Two in *Handbook for Ethnic Conflicts: International Perspectives, International and Cultural Psychology.* Berlin: Springer.

Sato, C. J. (1993). Language Change in a Creole Continuum: Decreolization? In K. Hyltenstam & Å. Viberg (Eds.), *Progression & Regression in Language: Sociocultural, Neuropsychological & Lingusitic Perspectives* (pp. 122–143). Cambridge: Cambridge University Press.

Schmitt, R. C., & Nordyke, E. (2001). Death in Hawaii: The Epidemics of 1848–1849. *The Hawaiian Journal of History, 35,* 1–13.

Shanks, G. D. (2016). Lethality of First Contact Dysentery Epidemics on Pacific Islands. *American Journal Medical Hygiene, 95*(2), 273–277.

Silva, N. K. (2004). *Aloha Betrayed: Native Hawaiian Resistance to American Colonialism.* Durham and London: Duke University Press.

Speakman, C. E., & Hackler, R. E. A. (1989). Vancouver in Hawaii. *The Hawaii Journal of History, 23,* 33–65.

Stannard, D. E. (1989). *Before the Horror: The Population of Hawai'i on the Eve of Western Contact.* Honolulu: Social Science Research Institute, University of Hawai'i.

Stannard, D. E. (1990). Disease and Infertility: A New Look at the Demographic Collapse of Native Populations in the Wake of Western Contact. *Journal of American Studies, 24,* 325–350. Reprinted in *Biological Consequence of the European Expansion, 1450–1800.* K. F. Kiple and S. V. Beck (Eds.). London: Ashgate.

Stannard, D. E. (2000, March). The Hawaiians: Health, Justice, and Sovereignty. *Cultural Survival Quarterly, 24,* 28–33.

Stokes, J. F. G. (1931, October 27). *Iron with the Early Hawaiians* (Papers of the Hawaiian Historical Society, Number 18. Papers Read Before the Society). Honolulu: Printshop.

Stover, J. S. (1997). *The Legacy of the 1848 Mahele and Kuleana Act of 1859: A Case Study of the Lāʻie Wai and Lāʻie Maloʻo Ahupuaʻa, 1846–1930* (A thesis submitted to the Graduate Division of the University of Hawaii in partial fulfillment of the requirements for the degree of Master of Arts in Pacific Islands Studies). University of Hawaii at Manoa, Honolulu.

Tcherkezoff, S. (2008). *First Contacts in Polynesia.* Canberra: ANU Press. JSTOR. Available at %20http:/www.jstor.org/stable/j.ctt24h2mx.5.

Townsend, E. (1888). Extract from the Diary of Ebeneezer Townsend, Jr. Supercargo of the Sealing Ship "Neptune" on Her Voyage to the South Pacific and Canton. As published in *Papers of the New Haven Historical Society* (Vol. 4). Hawaiian Historical Society Reprints (Number 4). Honolulu. Available at https://evols.library.manoa.hawaii.edu/bitstream/10524/642/1/RP04.pdf.

Tuhiwai-Smith, L. (1999). *Decolonizing Methodologies: Research and Indigenous Peoples.* Dunedin: University of Otago Press.

Uchiyama, M. (2016). *The Haumāna Hula Handbook for Students of Hawaiian Dance.* Society for the Study of Native Arts and Sciences. Berkeley: North Atlantic Books.

Vancouver, G. (1798). *A Voyage of Discovery to the North Pacific Ocean and Round the World, 1790–1795* (Vol. I, pp. 158–160, 187–188). London: Pall-Mall. Printed for G. G. and J. Robinson, Paternoster Row, and J. Edwards.

Weber, M. (1930). *The Protestant Ethic and the Spirit of Capitalism.* London and Boston: Unwin Hyman Publishers. First published in German in 1905 (T. Parsons, Trans.).

Yokum, C. F. (1967). Ecology of Feral Goats in Haleakala National Park, Maui, Hawaii. *American Midland Naturalist Journal, 77,* 418–451.

3

Traditional Use and Management of Natural Resources in the Hawaiian Islands

3.1 Overview

As with the previous chapters, the discussion that follows provides necessary context for Chapters 4 and 5, which summarize the long series of 'aha (meetings) or puwalu (people working together) first convened in 2006 with the assistance of the Western Pacific Regional Fishery Management Council, Native Hawaiian organizations, and various state agencies. The initial puwalu and those that followed enabled Native Hawaiians and other local residents to deliberate on the traditional use and management of natural resources and the role of tradition in contemporary and future natural resource policy-making processes across the islands. The meetings incorporated the 'ike (knowledge) and na'auao (wisdom) of hundreds of kāhuna mahi'ai'ana (expert farmers), kāhuna lawai'a (expert fishermen), and many other specialists, and thereby functioned to give voice to living kūpuna and ancestors who long ago pursued, used, and cared for Hawaii's natural resources.

Key points made in this chapter are that continually evolving traditional ecological knowledge, customs, and methods for pursuing and managing natural resources in the present-day have their molekumu

© The Author(s) 2019   **99**
E. W. Glazier, *Tradition-Based Natural Resource Management*,
Palgrave Studies in Natural Resource Management,
https://doi.org/10.1007/978-3-030-14842-3_3

(roots, origins) in the deep past. While the strength of these roots was tested in the nineteenth and twentieth centuries, and continue to be tested today, linkages to the past were by no means severed. Strong and persistent resistance to detrimental forms of exogenous change, new venues for documenting tradition, the persistence and recent expansion in use of the Hawaiian language, and the strong tendency of Kānaka Maoli to consult with living kūpuna and the words of the ancestors have continually vivified Hawaiian culture as a whole. In the realm of indigenous management of natural resources, this lends additional significance to the past and its role in shaping present and future deliberations and decisions about how the resources of land and sea may best be used and conserved in the face of inevitable social and environmental changes.

3.2   Oppression and Indigenous Resistance into the Twentieth Century

From the time of their first arrival in Polynesia, the Puritan evangelists sought to alter indigenous lifeways deemed contrary to appropriate Christian behavior. Such was the nature of the "mission" to be conducted among "a heathen people" in "New Zealand, Tonga … and the Society, Marquesan, Hervey, Friendly, Samoan, Feejeean [sic], New Hebrides, Hawaiian, and Micronesian" islands (Damon 1869: 7). Given the long tenure and sophisticated nature of Pacific island societies and cultures, and the condescending manner of the new arrivals, the initial interface rarely went smoothly.

Inducement of social unrest, introduction of disease, acquisition of land and resources, and oppression of indigenous lifeways were universal themes of externally sourced change across Oceania during the eighteenth and nineteenth centuries. The impacts to indigenous island societies were highly disruptive and persistent over time. Such societies all too often became divided in political terms, as certain leaders and residents espoused foreign ways and ideas while others chose the path of resistance and adherence to that which was familiar and trusted.

The implications of capitulating decisions made by indigenous leaders during this tumultuous period extend to the present day. Again, it was the powerful effect of introduced disease and mass death that led certain leaders and many makaʻāinana to question that which was known and to consider acceptance of the moral proclamations made by the missionaries. The missionary promise of redemption notwithstanding, loss of life and great suffering continued unabated among Hawaiian and other island societies of Polynesia during the 18th and 19th centuries.

In the case of Hawaiʻi, the proponents of the new religion were resourceful and determined. Once the missionaries gained a foothold, hula and key elements of ancient Hawaiian religious practices were banned by church edict (Silva 2004a: 88), as were the plant-resource dependent medicinal practices of the kāhuna lapaʻau (Donlin 2010: 217). Upon conversion to Christianity, certain aliʻi acted as they saw fit to integrate new moral strictures into kumukānāwai (code of law). In 1830, for example, the recently converted Queen Kaʻahumanu issued a verbal ban on the public performance of hula and forbade "the worship of ancient gods and all untrue gods" (Silva 2000: 30). Similarly, aspects of lapaʻau, such as ʻanaʻana (psychospiritual methods of healing) were banned in the 1860s and 1870s when the monarchy provided for government-issued board licenses that allowed indigenous medical experts to practice only herbal medicine and lomilomi (traditional massage) (Blaisdell 1996). Subsequent to the illegal overthrow of the Hawaiian kingdom—as arranged by American oligarchs in 1893—even the mellifluous Hawaiian language was suppressed through a ban on its use in public and private school instruction (Townsend 2014: 4; Kupau 2004).

As discussed above and in the previous chapters, the arrival of foreigners in Polynesia generated lasting challenges for the indigenous peoples of the Pacific islands. Yet, with the exception of that which the islanders found useful in practical-material terms, there was no immediate or consistent embracing of externally induced change. As was the case throughout Polynesia, the original settlers of the Hawaiian Islands consistently resisted that which they found undesirable, and had it not been for the introduction of deadly disease, Kānaka Maoli may well have deflected the evangelist and capitalist incursions of the haoles. As it was, rapid population loss and fundamental disruption to age-old systems of land tenure, food production, and sociopolitical organization required the descendants of Hawai'i's original inhabitants to engage a struggle for cultural survival into the twentieth century.

Resistance to Macro-Social and Economic Change

While concessions did occur, few Hawaiian leaders willingly or knowingly prioritized foreign political or economic interests above the well-being of the maka'āinana. When concessions did occur, they typically were strategic in nature, and most detrimental outcomes were unintended. For example, King Kalakaua's controversial signing of the Reciprocity Treaty of 1875, which allowed American firms tariff-free access to the sugar market and a desirable port of entry at Pearl Harbor, ultimately empowered the overthrow and annexation. But this was an unintended outcome of the king's interest in improving the economic status of the kingdom, which had become mired in debt during his reign (Silva 2004a: 89). According to Kuykendall (1967: 83), Kalakaua's support for the Reciprocity Treaty ultimately led to a major increase in sugar exports, from 24,566,611 pounds in 1874 to 330,822,879 pounds in 1890, and increase in total export values during the period.

It is reiterated here that Hawaiians as a whole never capitulated to the cultural or sociopolitical changes that were to alter life in the islands so profoundly during the twentieth century. The processes that led to the illegal overthrow of the Hawaiian Kingdom in 1893, and the overthrow itself, were strongly resisted by the ali'i then in power. Queen Lili'uokalani diligently sought to maintain the sovereignty of the Hawaiian people before, during, and after the coup d'état, and was imprisoned for eight months following an attempt by royalists to reinstate her to power in 1895.

Similarly, the illegal annexation of the Hawaiian Islands by the United States in 1898 was keenly resisted by the monarchy and by Kānaka Maoli across the islands (see Silva 2004b: 10). At the same time, many maka'āinana—increasingly bereft of family members, land, and traditional modes of social organization—continued to resist localized agents and processes of change as they had throughout the nineteenth century. Such resistance is evident in many accounts of the era.

Resistance to Strictures on Hula

As noted above, hula was denounced by the missionaries and verbally banned by royal decree. But evidence suggests that the dance form was merely forced underground and continued to be practiced with the support of chiefs, who were sufficiently bold to ignore the proclamation. According to Hong (2013: 11), who cites Tabrah (1984), "many chiefs ignored the edict and some secretly financed clandestine hula schools where the hula continued to be taught and practiced."

Hula was practiced in secret and even publicly after Queen Ka'ahumanu's ban of 1830, ultimately leading the missionaries to advocate for a formal constitution and written legislation that would undo such immoral behavior and "laxity" in adhering to the verbal proclamation. However, Silva (2004a: 36–39) has determined that the underlying intent of the missionaries in urging the monarchy to draft a formal constitution (and by extension to establish written kumukānāwai formally banning hula and other cultural practices) was related more directly to their own political and economic interests in Hawai'i than to concerns about declining morality. In any event, no written law forbidding hula was enacted after Ka'ahumanu's death in 1832, and hula continued to be practiced during the reign of Kamehameha III (Kauikeaouli).

There was outcry among Kānaka Maoli when public displays of hula were deemed illegal in the absence of proper licensing as per legislation passed in 1859 during the reign of Kamehameha IV (Alexander Liholiho). Of the legislation (which stopped short of a formal ban), concurrent public dissatisfaction with strictures on ancient cultural practices, and ongoing resistance to the suppression of culture, Silva (2000: 42) writes:

This [legislation] seems to have put an end to public discussion of hula for several years. It also indicates the strength of the quiet resistance: were the government to actually put forth a [full] ban, the *Polynesian* [a missionary-operated paper active in Hawai'i during the mid-1800s] expected the people to fight back. In the meantime, resistance to the deliberate destruction of Kanaka traditional culture arose in the form of a newspaper, *Kahoku o Ka Pakipika*, which printed traditional mo'olelo and mele and was edited by the future King Kalākaua. (Silva 2000: 42)

Resistance to Constraints on Lā'au Lapa'au

Irrespective of verbal and written laws to the contrary, lā'au lapa'au (medicine) continued to be practiced and modified throughout the nineteenth century and remains an important part of Hawaiian culture to the present day. Plant resources available in the islands were and remain essential to the practice. As discussed by Donlin (2010), resistance to suppression of traditional Hawaiian medicine was both adaptive and persistent:

Missionaries and their converts miscategorized all kāhuna, even medicinal kāhuna, as black magic sorcerers, or kāhuna 'anā'anā (Judd 1998:239) and called their work "evil" (ibid., p. 239). Foreign physicians living in Hawai'i also helped alienate the kāhuna by condemning the practices of their Native Hawaiian counterparts, although Western medicine at the time included practices such as bloodletting, leeching, surgery without anesthesia, and cauterization of severed flesh with burning-hot irons (Bushnell 1993:95)…Within several years after the arrival of the missionaries, medicinal kāhuna practices were outlawed (Kamakau 1992:307-308, 322). Kāhuna still practiced in secret in spite of being marginalized by the increasing dominance of Western culture and the demonization of their arts (Bushnell 1993:19). Their practices adjusted to the influx of Western influence, as many kāhuna adopted the use of non-native or Polynesian-introduced plants into their healing arts and called upon the Christian God in their prayers (Chan 1994:9). [Nevertheless] Kāhuna continued to practice their healing arts underground, until the Hawaiian government enacted new laws later in the nineteenth century that enabled some of them to emerge (Chun 1989:13). (p. 217)

A licensure process established in 1865 is said to have created both opportunity and cultural obstacles for kāhuna lapaʻau, some of whom resided in rural areas and/or could not effectively interact with the new bureaucracy. A Hawaiian Board of Health was established in 1869 to examine applicants' fitness to practice traditional medicine, but between 1873 and 1878, only fourteen Hawaiians were given license to practice (Donlin 2010: 210).

During the reign of King Kalākaua (1874–1891), the monarchy undertook a variety of measures to revive formerly prohibited aspects of Hawaiian traditions and culture, including traditional Hawaiian medicine, which often involved use of plant resources. Some 300 kāhuna reportedly became licensed during this era. But as related by Donlin (2010: 218), when the Kingdom was overthrown in 1893, the new haole republic repealed many laws, and kāhuna were once again forced into the background. As stated by Donlin (2010: 218) who cites Chun (2009a: 4), "as [was the case] in the period between the outlaw of kāhuna after the missionaries arrived and [when licensing was enabled], this [the overthrow] once again made kāhuna's practices forbidden and forced them to work in secrecy."

A law enacted in 1919 allowed traditional healers to once again become licensed and practice openly. But in the 1940s, the Territorial Board of Health established a Board of Examiners requiring that applicants test for licensure. Few kāhuna were given license, however, in part because of lack of understanding about Hawaiian culture and medicine on the part of the primarily haole certifying board (Donlin 2010: 219).

When Hawaiʻi became a state in 1959, the 1919 traditional healing law was deemed obsolete, and by 1965, only lomilomi (traditional massage) could be lagally practiced in Hawaiʻi, and only under the state's (contemporary) massage licensing law. As one crucial element of the Hawaiian Renaissance, initiated in the mid-1970s, the kāhuna lapaʻau (traditional medical expert), who had continued to practice the full range of techniques primarily within ʻohana and local community settings around the islands irrespective of haole legalities, began to consider means for expanding their kuleana and formally preserving the ancient healing traditions through yet another formal arrangement (Donlin 2010: 224 cites Papa Ola Lokahi 2008).

After a series of discussions within the community of healers, and external implementation of various legislative strategies, Act 162 (aka the Healer's Law) was passed in 1998. This exempted Hawaiian healing arts from a generalized statewide prohibition on unlicensed medical practices. The rationale for the legislation involved the clear desirability of traditional methods that could improve the health of Native Hawaiians, and "an urgent need to keep alive the healing arts, which were at risk of fading away with the elderly kāhuna, who were perishing at a rapid pace" (Donlin 2010: 213). As is common when indigenous people seek to guard tradition from oppressive political agendas of colonial powers, the integration of Hawaiian healing traditions into the formal medical-legal framework required extensive deliberation among the practitioners, and persistent interaction with legislators and agencies over many years.

Resistance to Suppression of the Hawaiian Language

Although the colonial powers of the late nineteenth century sought to stifle Hawaiian culture by suppressing its linguistic expression—a tactic similarly employed to suppress Alaska Native and American Indian cultures on the continent (e.g., see Stout 2012), in reality ultimately nothing could prevent the continued use of Hawaiian. Despite the post-overthrow ban on the use of Hawaiian in schools around the islands (Act 57, Section 30 of the 1896 Laws of the Republic of Hawai'i), the language continued to be spoken in many homes and community settings, and it was used in written form in numerous Hawaiian language newspapers through at least 1948 (Silva 2004a). Moreover, it has always been spoken as the primary language on the island of Ni'ihau (see Ni'ihau Cultural Heritage Foundation 2018). It is important to note that the language is critically important among indigenous persons who transmit subtle cultural knowledge of natural resources across generations.

Based on a timeline of the life of the Hawaiian language provided by Pūnana Leo (2018), a sort of linguistic renaissance was initiated in Hawai'i in 1915, wherein use of the language was prioritized in Hawaiian churches, civic clubs, and other public venues attended by Kānaka Maoli. By 1919, the territorial legislature had enacted laws requiring that Hawaiian be taught in public high schools, in the Territorial teacher preparation program, and in elementary schools serving children in Hawaiian Homestead areas ('Aha Pūnana Leo 2018). Readers should note that Hawaiian homesteads were established through the Hawaiian Homes Commission Act of 1920, the intent of which was to enhance "economic self-sufficiency" by providing lands on which displaced Hawaiians could reside, cultivate food, and raise livestock through long-term lease arrangements (see Department of Hawaiian Home Lands 2018). Many Native Hawaiians continue to reside on such lands and many others continue to wait for lease arrangements to be administered in this twenty-first century.

The work of Mary Kawena Pukui (1895–1986), who translated Hawaiian newspapers and documented traditional life in Hawaiian communities with Martha Beckwith during the mid-twentieth century, and who completed and published the *Hawaiian-English Dictionary* with Samuel Elbert in 1964, was instrumental in generating an accurate record of the ancestral language for use by Hawaiians and others interested in this ancient Polynesian culture. Subsequent efforts by Hawaiians such as Dorothy Kahananui, Larry Kimura, Edith Kanaka'ole, Ilei Beniamina, John Waihe'e, Kauanoe Kamana, Pila Wilson, and many others were similarly instrumental in preserving and advancing the language during the twentieth century.

Hawaiian language programs were ultimately established by the University of Hawai'i in the early 1980s, and legal barriers to the functioning of privately operated Hawaiian language immersion schools known as Pūnana Leo (Nest of Voices) were overcome in 1986. In 1999, for the first time in over a century, a cohort of K-12 students educated entirely in Hawaiian graduated from schools in various parts of the archipelago (see 'Aha Pūnana Leo 2018). The Hawaiian Language Immersion Program *Ka Papahana Kaiapuni* was successful. Notably, the founders of the standing council known as 'Aha Pūnana Leo, in conjunction with U.S. Senator Daniel Inouye (D-Hawaii') and various tribal leaders on the continent, were instrumental in the passage of the Native American Languages Act of 1990, which preserves, protects, and promotes "the rights and freedom of Native Americans [including Alaska Natives, Native Hawaiians, and other Pacific Islanders] to use, practice, and develop" their native languages.

Indicative of both resistance to pressures that would oppress the Hawaiian people and the success of efforts to revitalize the language and the ancient culture it expresses, Bushnell (1986: xi) describes the status of ʻōlelo ʻōiwi (the language of the sons of Hawaiʻi) roughly a decade into the renaissance of the 1970s. A language once threatened now persists in the daily discourse of many islanders (Bushnell 1986):

> The language lives, with an astonishing resilience. It graces the conversations of older folk and not only those who dwell in Kona or Niʻihau; it is heard in the prayers uttered at weddings, funerals, family gatherings, meetings of clubs and societies, and dedications of buildings, offices, restaurants, and highways. It is spoken in fishing villages and cattle ranches. It quickens the songs and chants accompanying dances and rituals, whether solemn or profane. And it wells up daily in our speech and thoughts, almost in unconscious response to our island heritage. (p. xi)

3.3 The Persistence of Traditional Pursuit, Use, and Management of Natural Resources

The practice of hula and traditional medicine and use of the Hawaiian language were jeopardized by outside sources of change and even by those aliʻi who were strongly influenced by foreign people and ideas during the nineteenth century. But each of these vitally important dimensions of Hawaiian culture ultimately survived through popular resistance and ongoing practice in secret. Knowledge developed over many centuries was not readily abandoned. On the contrary, the persistent attention of Kānaka Maoli to their ancient molekumu (origins), their ʻōlelo (language), and the wisdom and knowledge of their kūpuna, including traditional knowledge of the natural world, have kept a Polynesian society and culture alive and distinctive to this day.

Of the various cultural practices that sustained Hawaiians over the centuries, the harvest, use, and management of natural resources have been most directly essential to survival. Indeed, as the clouds of introduced disease and other imposed changes darkened Hawaiian society during the nineteenth century, the need to wrest a living from land and sea became ever more critical. Any attempt at suppressing this realm of culture was futile. Hawaiians resisted changes to crucial dimensions of their culture even as they were increasingly displaced from their ancestral lands and enmeshed in a cash economy. People retained social access to the resources of land and sea, and the practices of lawa'ia, and mahi'ai'ana (farming) continued to evolve and remain central to survival.

Many Hawaiians were able to persist on the 'āina, if only as tenants. Individuals and 'ohana forced into Hawai'i's growing towns and cities tended to maintain strong relations with those who were able to stay on the land. Supplementation of wage-based household economies with subsistence foods, barter, gifting, customary exchange, remittances, and generalized and specific reciprocity were characteristic of Native Hawaiian life during the Plantation era and beyond. With the exception of the most populous island areas, both the land and ocean remained productive sources of food well into the twentieth century.

In reiteration, as social and economic conditions deteriorated after the arrival of the haole, the indigenous people of the Hawaiian Islands depended increasingly on a nuanced understanding of their natural surroundings; on well-tested methods of fishing, hunting, gathering, and farming; and on inter-generational knowledge about how best to care for that which could feed them in the worst of times. Despite the tragic loss of people and pressure to abandon traditional lifeways, core cultural practices were neither abandoned nor successfully oppressed.

Significantly, traditional ecological knowledge, transmitted across generations, was given an additional venue for documentation and widespread expression when the missionaries enabled the development of written language, and when Hawaiian language newspapers such as *Ka Hoku Pakipika*, *Ka Nupepa Kuakoa*, and many others printed mo'olelo focused on various aspects of Hawaiian culture during the nineteenth and twentieth centuries. News features and culturally oriented articles described ancient traditions and addressed issues related to the ongoing use of marine and terrestrial resources among the indigenous islanders. Today, such information is increasingly mined by Native Hawaiian and other scholars who retain an interest in Hawaiian history and its implications for present and future generations.

Mahi'ai'ana

The forms of traditional ecological knowledge that were developed and used by generations of Hawaiians are as varied as the region's natural resources and the terrestrial and ocean ecosystems that support them. As discussed throughout this book, the kūpuna were adept farmers, fishermen, and gatherers of food and essential materials from the natural world—sufficiently adept to enable a small population of ocean voyagers to evolve into a complex and burgeoning island society. Fishing, farming, and associated knowledge of island and ocean were developed and refined over many centuries, ultimately constituting an evolving base of ecological and practical knowledge commensurate with or surpassing any in the Pacific islands (Fig. 3.1). Of the agricultural accomplishments made by the ancient Hawaiians, Handy et al. (1972: vii) write:

> In their practice of agriculture, the ancient planters had transformed the face of their land by converting flatlands and gentle slopes to terraced areas where water was brought for irrigation by means of ditches from mountain streams. The making of terraces and ditches and their maintenance, and the regulation of water, entailed much cooperative and communal labor organized under [the konohiki]. In all the Polynesian islands there was some organized work in some direction in canoe building, house construction, fishing, preparation of food on a large scale for feasts, war-making, and other communal activities. But there was nowhere the continuous organized enterprise comparable to that which was essential to the systematic gardening operations of Hawai'i.

The work of Noa Kekuewa Lincoln and Peter Vitousek (2017) supports the perspective that agriculture reached a level of production unrivaled in Polynesia. Based on extensive paleo-agricultural data from the Kohala

Fig. 3.1 Molokaʻi woman weaving lau hala (pandanus) mat in 1913 (Photograph by Ray Jerome Baker)

District of Hawaiʻi Island, the authors assert the effectiveness of the field systems used in antiquity. Kohala is a particularly dry region at lower altitudes, and thus extensive lava rock wall infrastructure had to be constructed to move water from upslope areas in order to nurture the fields below, where thousands of acres of ʻuala (sweet potatoes; *Ipomoea batatas*) and other vital crops were grown. As such, the ancestors "created intensive rainfed agricultural systems that were unique in Polynesia," and which ultimately fostered development of powerful chiefdoms, including that of Kamehameha I (Lincoln and Vitousek 2017: 1).

In the prologue to "Indigenous Polynesian Agriculture in Hawaii" (Lincoln and Vitousek 2017), Lincoln describes his astonishment at the scope and scale of what his kūpuna had accomplished in the ahupuaʻa of Ka Mokuʻo Kohala, and by extension elsewhere in the Hawaiian Islands:

> I walk through a cattle pasture in the North Kohala district of Hawaiʻi Island...As we trudge up the steep hill, battling 35 mile-an-hour trade

winds, I glance at the sparse ironwood trees bent over by the wind and wonder how anything productive could grow in this harsh environment. Finally, after reaching the summit, I pause to catch my breath [but] … what I see takes my breath away. Where moments ago there was nothing but a grassy expanse, the elevated perspective reveals, like a magic 3-D eye book, row after row after row of ancient agricultural infrastructure extending as far as can be seen, off into the horizon. These are the field walls that formed the backbone of Hawaiian dryland agriculture…Such a feat of landscape alteration could only have been accomplished by a massive labor force within a highly organized society. This was not the works of villages and family clans, or small chiefly hierarchies. This was the work of a nation.…We continue to uncover some of the adaptations, innovations, and mechanisms used to achieve an incredible scale of food production—one that sustained a population greater than exists on most of the Hawaiian Islands today and was accomplished without the use of any external inputs. Our aim now is not only to use this information to better understand the past, but to apply the lessons learned from our ancestors to inform and help guide modern agriculture development in Hawai'i today. (p. 2)

Kalo (taro; *Colocasia esculenta*), the source of poi (a starchy paste made from cooked taro corms that have been pounded and thinned with water), 'uala, 'ulu (breadfruit), mai'a (plantains and bananas), and various other crops and fruits were, and in some 'ohana around the islands remain, essential sources of food for Native Hawaiians. Kalo itself includes many varietals that are of value to Kānaka Maoli. Lincoln and Vitousek (2017: 8) list 30 key crops used by Hawaiians of antiquity for food, medicine, shelter, oiling, poisoning, and other vital purposes, most of which were brought to the islands by the early Polynesian colonists (see MacCaughey 1917).

The labor and expertise of Hawaiians in working and managing the land to enable consistent production of food and other resources were pivotal to the early survival and ultimate blossoming of a complex Polynesian society. As described by Andrade (2008), even the *Kumulipo*, the cosmological genealogy that explains the origins of the Hawaiian universe, prioritizes the role of the lā'au (plant) in giving birth to and sustaining Kānaka Maoli from time before history:

The narrative begins with the birth of the earth out of Pō, the time of eternal darkness, and progresses through many eras, eventually portraying all aspects of the cosmos and all beings inhabiting it. This story celebrates the union of earth mother (Papa) and sky father (Wākea) from which islands are born. Also born was the kalo (taro plant), named Hāloanakalaukapalili (Hāloa of the trembling leaf), followed by the earliest human ancestor of the Hawaiian people, Hāloakanaka (Hāloa the human being). (pp. 4–5)

Today, Kānaka Maoli are largely displaced from the agricultural potential of the land. Kamehameha Schools and the Department of Hawaiian Homelands do oversee large tracts of land around the islands, with ongoing and potential benefits for the Hawaiian people. But the seminal actions of the missionaries and entrepreneurs of the nineteenth century to institute private ownership of land has gradually led to the contemporary situation, wherein vast acreages on all the islands are held and/or used by: the federal government; the State of Hawai'i; large corporations such as Alexander and Baldwin, Maui Land & Pineapple, and GL Limited (among many others); and many individuals whose ancestors are not Polynesian.

Meanwhile, the vast majority of land in urban zones and surrounding areas has been divided and sub-divided many times. Hawaiians and part-Hawaiians now work the 'āina primarily on relatively small and often disparate rural parcels around the main islands, for purposes of subsistence and, increasingly, for commercial sale of food products. But pollution associated with Plantation era pesticide use is widespread on the land (e.g., see Churchill and Venue 2004: 707–708) and, as noted by Melrose et al. 2016: 6), "the increasing value of Hawaii's real estate has a significant impact on farmer's ability to affordably acquire and farm land...Productive farm areas like Kula Maui, Kilauea Kauai, and South Hilo are undercut by increased real estate prices and by new owners without agricultural intentions." The years of coherent and highly functional ahupua'a are, at this point in time, a chapter in history, a situation that is in no way lost on those who continue to seriously question the legality of past and present actions and processes that have led to the current status of land and land use across the archipelago.

Lawaiʻa

The pursuit and use of highly nutritious protein-laden seafood were as important to the survival of the Kanaka Maoli over the centuries as were the practices and products of mahiʻaiʻana. Given the seafaring nature of the original colonists of Hawaiʻi, indeed of all Polynesians, fishing knowledge and skills have their origins in deep antiquity. While the Polynesian voyagers who travelled the vast Pacific transported a variety of land-based foods from island to island, it is quite obvious that fishing skills and seafood helped sustained them along the way. As noted by Holmes (1993: 12) of the ancient voyages, "undoubtedly there were trolling lines out all day, every day."

The many skills and techniques exhibited by Hawaiian fishermen during the post-contact and Plantation eras are well-documented in the literature of the era. Kamakau (1976: 60), for instance, writes that that "there were as many types of fishing as there were fish and that Hawaiians were adept at all of them." The author goes on to say that "as fishing was done by the ka poʻe kahiko (assembly of ancients) so it is done now [in the nineteenth century]—it is impossible to improve on their methods."

The scope and proficiency of fishing activities among Hawaiians of old is truly notable, as is the persistence of the knowledge and skills involved. Even today, fishing, seafood, and knowledge of the marine environment remain central to the practice and expression of Hawaiian culture (e.g., Allen 2013; Geslani et al. 2012; Glazier and Kittinger 2012; Glazier et al. 2013; Kittinger et al. 2015; Maly and Maly 2003 I and II; McGregor 2007; NOAA Fisheries Pacific Islands Fisheries Science Center 2018; Vaughan and Ayers 2016; Vaughan and Vitousek 2013).

Although modern gear and vessels have now generally superseded those used in the nineteenth century and deeper past, even this statement should be qualified. For example, outrigger canoes, both motorized and paddled, are used today to pursue various reef, nearshore, benthic, and pelagic species along parts of Hawaiʻi Island. Moreover, many fishermen utilize traditional ecological knowledge and ancient fishing methods and strategies in conjunction with modern vessels and gear. Some even use modern versions of ancient gear, such as that required to implement the ancient palu ʻahi method for capturing tuna with handlines. Finally, a range of practical-traditional fishery management approaches continue to be used to care for marine and coastal ecosystems and associated resources around the islands. Given localized variability in sociocultural and ecological conditions, the focus of these approaches often differ in form and function by island, island district, and even specific ahupuaʻa. For instance, indigenous residents of Maui traditionally used nets in bays and shallows to a great extent, and therefore nets were the focus of various kanāwai and kapu. Similarly, Molokai residents were and continue to be dependent on reef fish, and thus the methods used for capture of such species were the principal focus of local rules and restrictions (Kaʻaiʻai 2018).

Variability aside, general principles tend to guide collective treatment of natural resources by indigenous Hawaiians. This is expressed by Poepoe et al. (2003) in their discussion of an ongoing culturally-based approach to the harvest and management of marine resources that has roots in deep antiquity along a remote coastal zone of Molokaʻi:

> It is traditional for Hawaiians to "consult nature" so that fishing is practiced at times and places, and with gear that causes minimum disruption of natural biological and ecological processes. The Hoʻolehua Hawaiian Homestead continues this tradition in and around Moʻomomi Bay on the northwest coast of the island of Molokaʻi. This community relies heavily on inshore marine resources for subsistence and consequently, has an intimate knowledge of these resources. The shared knowledge, beliefs, and values of the community are culturally channeled to promote proper fishing behavior. This informal system brings more knowledge, experience, and moral commitment to fishery conservation than centralized government management. Community-based management in the Moʻomomi area involves observational processes and problem-solving strategies for the purpose of conservation. The system is not articulated in the manner of Western science but relies instead on mental models. These models foster a practical understanding of local inshore resource dynamics by the fishing community and, thus, lend credibility to unwritten

standards for fishing conduct. The "code of conduct" is concerned with how people fish rather than how much they catch. (p. i)

In sum, traditional Hawaiian concepts and knowledge regarding the behavior and suitability of fish for consumption, important attributes of marine ecosystems, and culturally appropriate means for pursuing, using, and managing marine fisheries and marine resources are being applied around the islands in this new century (Jokiel et al. 2011; Glazier 2011). Many challenges and potential solutions for effectively managing such resources are discussed at length in the following chapter of this book.

Traditional knowledge of marine resources provides an excellent topic through which to examine natural resource use and management over time and into the present era in the Hawaiian Islands. The scope of such knowledge and the nature and extent of indigenous fishing practices is indicated in a variety of sources, including *Ka 'Oihana Lawai'ia— Hawaiian Fishing Traditions* (2006), a descriptive compilation of fishing methods and related mo'olelo based on the work of Daniel Kaha'ūlelio (1835–1907), an expert fisherman, teacher, lawyer, and public servant who published a detailed series of articles about traditional fishing in *Ka Nupepa Kuakoa* after the turn of the century.

Kaha'ūlelio hailed from a long line of fishermen whose knowledge and experiences extended back to a time when the ahupua'a system was fully intact and Hawaiians had not yet encountered the haole explorer, missionary, or entrepreneur. The mokuna (chapter) titles provided in Puakea Nogelmeier's (2006) translated compilation of Kaha'ūlelio's articles are of themselves indicative of the expansive nature of the topic and the (partial) scope of traditional knowledge possessed by the kāhuna lawa'ia (fishing experts) of old Hawai'i (see glossary for basic definitions of listed fish, gear, and methods):

- Lau Nui Fishing;
- Hi Aku Fishing;
- Fishing in the Open Ocean;
- Fishing with Octopus with a Cowry Shell Lure;
- 'Ōkilo Fishing for Octopus; Octopus Spearing;

- Fishing for Schools of Mālolo, Iheihe, and Puhikiʻi;
- ʻŌpelu Fishing;
- Fishing with Melomelo;
- Fishing with Palu; Kāʻili Fishing;
- Kumu and ʻAhuluhulu Fishing;
- Ōʻio Fishing;
- Uhu Fishing;
- Fishing for Ulu Kaʻi;
- Kolo Net Fishing; Hōauʻau Fishing;
- Hoʻomoemoe Fishing;
- Kuʻikuʻi and Pahoe Fishing for Ulua;
- Fishing for ʻĀlalaua and ʻĀweoweo;
- Nenue Fishing;
- Hoʻoluʻuluʻu Hinālea Fishing;
- Diving for Weke;
- Akule Fishing;
- Paeaea Fishing;
- Kala Kū Fishing;
- Maomao Fishing;
- Mahimahi Fishing;
- A ʻuaʻu Fishing;
- Turtle Fishing;
- Fishing with a Spear;
- Luelue Net Fishing;
- Pōuouo Net Fishing;
- Fishing for Pond Mullet (ʻAnae);
- Uouoa Fishing;
- Pihā Fishing;
- Nehu Fishing;
- Eel Fishing;
- Torch Fishing;
- Kiolaula Fishing;
- ʻOpihi Fishing;
- Moi Fishing with a Net;
- Ka Lāʻau Fishing;
- Holoholo Fishing (pp. iv–vi).

In fact, as provided in the voluminous descriptive work of Kepa and Onaona Maly (Maly and Maly 2003), titled *Ka Hana Lawaiʻa a Me na Koʻa o na Kai ʻEwalu—A History of Fishing Practices and Marine Fisheries of the Hawaiian Islands*, the list above constitutes only a fraction of fishing methods and marine resource management approaches used by indigenous Hawaiians over the last couple of centuries. Documentation of lawaiʻa and related resource management strategies is thorough in this source, incorporating many scores of interviews with kahuna lawaiʻa who had engaged extensively in fishing and shoreline gathering activities during the twentieth century. The descriptions are so detailed that there is risk to local ecosystems should contemporary fishermen re-adopt highly efficient methods of the past without also following restrictions formerly associated with their use. Notably, numerous persons consulted during the project hailed from families with an extensive history of involvement in local fisheries, and thus the report documents ʻike handed down from kūpuna who had lived and fished in various ahupuaʻa around the islands even during the nineteenth century. Finally, the work incorporates review of testimony gathered by Hawaiʻi's Boundary Commission, which was established in 1862 to (a) delineate components of ahupuaʻa and ʻili that had been awarded as private property during the mahele and (b) "identify knowledgeable native residents and kamaʻaina from whom detailed testimonies and descriptions of the lands and rights could be recorded" (Maly and Maly 2003 I: vii). Such information is of value to contemporary and future scholars, activists, and others who seek to understand events and actions occurring during a crucial time in Hawaiian history. Commission records include, but are not limited to, the following descriptions:

- Defined fishing boundaries; description of the place and type of fishery resources the area was noted for; limitations on the kinds of fish and who could take them; associated rituals; and choice fish held under kapu;
- Fish typically selected by the people, typically including: ʻahi, akule, amaʻama, malolo, manō, ʻoʻopu (from the mountains), ʻōpelu, uhu, and ulua;

- Method of fishing, including: ō (spearing); upena and kuʻuna (setting nets); makau and pā (hooks and lures); hāhā (trapping in one's hand); and loko (fish ponds, both natural and man-made);
- Loko, loko iʻa, kuapā, puʻu one (dune banked fish ponds); and ponds in which fish and kalo (loko iʻa kalo) that were grown together and named or identified on the islands of Hawaiʻi, Maui, Molokaʻi, Oʻahu, and Kauaʻi; and the presence and use of mākāhā (sluice gates);
- Referenced practices of: canoe making; preparation of nets and fishing line; collection of human bone for hooks; making paʻakai (salt), and the exchange of fish for other goods (Maly and Maly 2003 I: viii).

Numerous other sources of information are pertinent to documentation and analysis of traditional ecological knowledge, fishing methods, and the socioeconomic and cultural context of marine resource use and management in Hawaiʻi past and present. These include a wide range of books, Hawaiian newspapers, academic articles, archived correspondence, cultural resource management literature, and unpublished technical literature (e.g., see Newman 1970).

The following draws from a variety of such sources to provide the reader with basic understanding of traditional use and management of marine resources in the Hawaiian Islands. While many gears and strategies were and are used in the islands, the discussion focuses especially on open ocean fishing, culturally prized deep water bottomfish and pelagic species, and associated social and environmental context. Other fish and methods of capture are discussed, but open ocean fisheries are used to elaborate basic concepts about lawaʻia and the sociocultural context in which fishery management was and is now undertaken around the Hawaiian Islands.

Pelagic fisheries remain particularly important and popular in Hawaiʻi given: the cultural desirability of the tunas and other pelagic species at the dining table, close proximity to deep ocean waters, and a relatively high yield of meat per unit effort. Commercial operators focus especially on pelagic species, as do many non-commercial captains. Chan and Pan (2017: 2) report that an increasing number of small boat operators retain commercial marine licenses, with over 1800 such persons active in 2013, up from 1587 in 2003. Many more small-boat operators fish in the pelagic zone for recreation and production of food. Teneva et al. (2018: 31) assert that some 33% of Hawai'i landings are non-commercial in nature, and that some 90% of seafood landed in Hawai'i today is pelagic-sourced (p. 1), with particularly extensive landings by the region's longline and handline fleets. Of note, the region's longline fleet is regulated through limited entry permit and landings quota programs, and with provisions for protecting sea turtles and other species. But small-boat fishermen operating in the open sea, especially non-commercial operators, are subject to relatively few regulations. Many, *but certainly not all* contemporary small-boat operators therefore tend to self-regulate their activities with concern for traditional mores holding that one should take only what is needed during a given trip, and that spawning potential should be maximized by avoiding the harvest of juveniles. As noted in Chapter 5, tuna stocks are not presently considered overfished around the Hawaiian Islands (Sibert 2018).

Lawai'a Kahiko, Lawai'a Nupaikini

Polynesians were and remain a maritime people. Horticulture and animal husbandry were important for survival in their contribution to the "portable economy of animals" used during voyages and upon settlement on the various islands (Irwin 2006: 74–75). But Polynesians (and proto-Polynesians) traditionally dwelled primarily along the shorelines and spent extensive time fishing the reef, the nearshore zone, and ocean waters that often rapidly descend to great depths just offshore (Groube 1971: 312). As discussed by Kirch (1984):

> The sea provided the dominant source of protein in Polynesian diets. Littoral and inshore habitats yielded a variety of seaweeds, mollusks, holuthurians, sea urchins, octopus, squid, crustaceans, and small fish, while the deeper reefs, benthic, and pelagic zones produced larger fish,

rays, sharks, turtles, and cetaceans. Diversity and productivity in the marine components of island ecosystems [of Polynesia] vary considerably. For the most part, the central, tropical islands support extensive barrier reef and lagoon ecosystems with high organic productivity. Many of the subtropical islands have much less extensive reef development, and a less diverse marine fauna. Easter Island, lacking a fringing reef and isolated in the East Pacific, has only 140 species of fish, compared with about 450 species in Hawai'i, and more than 1000 species in Fiji. (p. 24)

The Polynesian mariners who reached the Hawaiian Archipelago did so in part because they were highly skilled fishermen. The acquisition of protein from fish and other marine resources was, therefore, a familiar pursuit and an essential aspect of the early subsistence and trade economies. Goto (1986: 378) asserts that the earliest evidence of fishing found to date in Hawai'i is at the Bellows site on O'ahu, where one- and two-piece fishhooks are found in conjunction with fish bones and mollusk shells. Evidence of fishing activity sustained over many centuries has also been recovered from sites in Hālawa Valley on Moloka'i, and at Ka Lae (literally "the Point," meaning South Point) in Moku o Kā'u on Hawai'i Island (Kirch 1974: 110). Upwelling and colliding currents continue to make Ka Lae a productive, if treacherous, fishing locale. A range of fishing activities occurred in this area in antiquity, evinced by the presence of one- and two-piece fishhooks, aku lure shanks and points, crescent points used on large wooden hooks for sharks and pelagic species, and various sinkers (Goto 1986).

Women were often responsible for the gathering of shoreline resources, including shellfish, 'opihi, limu, urchins, and crabs (Connors 2009: i). The nearshore zone was productive, and much fishing took place there—with hukilau (seine nets), various other nets, basket-like traps, spears, and pole and line. As noted previously, fishponds provided a consistent source of nutrition after the fourteenth century.

Fish bones are found in upland as well as coastal archaeological sites around the islands, indicating trade between residents within and across a given ahupua'a or moku. Notably, analysis of fish bones found at camp and village sites suggests that offshore food resources were generally pursued with less frequency than those in the nearshore and shoreline zones (Kirch 1985: 208). Goto (1986: 448) concurs and states that "it seems that Hawaiians generally preferred using inshore [near-shore/reef-associated] resources, where and when they were available, to using offshore resources." This may be due to relatively fewer fish at depth, the increased difficulty and risk of fishing in the offshore zone, a gap in the archaeological record, or a combination of these factors (Table 3.1).

But deep-water fishing certainly did occur, and open ocean fishes were much sought after for dietary, religious, and socioeconomic reasons (Goto 1986: 449–453)—as reflected in archaeological data and theory and in the Hawaiian language itself. According to Kirch (1985: 208–209) the deep-sea or pelagic zone of exploitation occurred beyond the benthic, which was between 30 and 350 meters in depth. The waters of the open ocean or dark blue sea were called kai pōpolohua mea a Kāne, as differentiated from the waters along the coast (lihi kai), the place before the waters become very dark (kai lū heʻe), and so on (Kamakau 1976: 11). Ocean waters were differentiated in other ways as well. For instance, a loud sea, i.e., one with tumultuous breakers, was called kai leo nui, literally "great voice of the sea" (Clark 2011: 246). For obvious reasons, fishermen using small craft avoid the shallows

Table 3.1 Marine resource harvest strategies in old Hawaiʻi[a]

Method	Harvest zone			
	Shoreline	Reef/inshore	Nearshore/benthic	Open ocean/pelagic
Gathering/gleaning	✓	✓	–	–
Poisoning	✓	✓	–	–
Snaring	–	✓	–	–
Spearing	–	✓	–	–
Traps	–	✓	–	–
Nets	–	✓	–	✓[b]
Octopus hooking	–	✓	✓	–
Shallow angling	–	✓	–	–
Deep angling	–	–	✓	–
Trolling	–	–	–	✓

[a]After Kirch (1985: 208)
[b]Malolo net

when kai leo nui is heard, as do those who normally traverse the shoreline on foot or dive in the nearshore.

As continues to be the case today, Hawaiian fishermen valued, sought, and captured various bottom fish species in the benthic zone, such as: ʻōpakapaka (pink snapper; *Pristipomoides filamentosus*), ehu (squirrelfish snapper; *Etelis carbunculus*), kalekale (Von Siebold's snapper; *Pristipomoides sieboldii*), and ulaʻula koae (Longtail red snapper, *Etelis coruscans*, now commonly known in Hawaiʻi as onaga), among many others. Highly valued pelagic species such as ʻahi (yellowfin tuna; *Thunnus albacares*), mahimahi (dolphinfish; *Coryphaena hippurus*), ono (wahoo; *Acanthocybium solandri*), and aku (skipjack tuna; *Katsuwonus pelamis*) were and continued to be avidly pursued and captured in the open ocean. Kūkaula (handline fishing) is undertaken in both zones, with trolling most consistently used in the open ocean.

The extent of pursuit of deep sea species has always varied across the islands, depending on practical considerations, such as the tendency of seasonal sea states to restrict availability of nearshore resources and ease of access to offshore waters. For instance, Goto (1986: 449–450) notes such differences between the Waikīkī and Waiʻanae areas on Oʻahu. Although both sites typically are protected from strong tradewinds and trade swell, the former is only periodically exposed to large Southern Hemisphere-origin swells (in summer), while the latter is frequently exposed to very large west and northwest swells (in winter). Reef fish species are abundant in the archaeological record at Waikīkī sites, where shoreline fishing activities were possible almost year round in part due to the swell-blocking effects of extensive fringing reef. Benthic and pelagic species were and are more commonly landed at Waiʻanae sites, where there is less expansive fringing reef, where deep water and productive bathymetric features relatively close to land offer good possibilities for pelagic fishing, and where deep channels afford some safety for mariners seeking to avoid breaking swells en route to the fishing grounds (Fig. 3.2).

The assertion that nearshore areas were favored for fishing may also have a political explanation. That is, ahupuaʻa boundaries and use of ocean resources were, in certain areas and points in history, contested inside the point at which breakers occur and highly regulated beyond. For instance, Campbell (1819) states that "the occupiers, or proprietors of land are entitled to the privilege of fishing upon their own shores as far as the tallest man in the island can wade at low water, and they may exercise that right at all seasons; but beyond that the sea is tabooed, except at two periods in the year, of six weeks each, during which unlimited fishing is allowed; at these times it is the general employment of the natives, and they cure enough to serve them through the tabooed season" (p. 138).

Thus, in some cases, if an ahupua'a did not have a protected near-shore zone and the konohiki of adjacent ahupua'a disallowed use by neighboring residents, it may have been necessary for fishermen to travel to the deep sea rather than seek shoreline or reef-associated resources in an adjacent ahupua'a. Ahupua'a boundaries and the kuleana of the konohiki could extend to fishing locations and resources in the offshore zone as well (Kosaki 1954 and Meller 1985). In some few areas around the islands, pelagic species occur particularly close to land—close enough to capture from the shoreline—such as along the leeward portion of Ka Lae on southernmost Hawai'i Island.

Fig. 3.2 View toward the deep offshore fishing grounds around Keawa'ula and Kaena Point, leeward O'ahu

The ongoing interest of Hawaiians in residing by the shoreline is both profound and practical. This point is emphasized by Handy et al. (1972) who, during the 1950s, lived, farmed, and fished with residents of what arguably is one of the most rugged regions in the island chain:

> Because of the wealth of food offshore, the 'ohana of Kama'oa [a rugged ahupua'a in Kā'u district and the location of Ka Lae] revered even these desolate seaward zones, ingeniously finding ways and places for cultivation, adapting themselves in some places to an almost desert existence in order to live close by the best fishing grounds. (p. 225)

Some contemporary Hawaiian 'ohana have for many generations managed to reside near the ocean in Hawai'i. But this is increasingly difficult in the current economic context. Coastal real estate is ever more valuable, tax rates increase annually, and even basic access to the shoreline is constrained by an increasing number of gates and fences. As a result of the mahele and shrewd acquisition of prime lands by haole entrepreneurs (Hasager and Kelly 2001), much of the land set aside for use by Native Hawaiians under the Hawaiian Homelands Commission Act was neither along the immediate coast nor highly favorable for agriculure.

Methods, Gear, Aku, and the Outrigger Canoe

Numerous scholars have examined fish hooks in their archaeological and functional contexts. Goto (1986) provides analysis of the topic with reference to prehistoric ecological and social conditions in Hawai'i. Stylistic and functional variety abound in the archaeological record. Some hooks were composites, others were constructed from single materials such as mammal bones (including whale bone), mother-of-pearl, cowry shell, wood, and ivory. Hooks made of human bones were not uncommon, and as Kirch writes (1985: 204), "the largest one-piece and two-piece hooks were made from human long bones [e.g., the femur and humerus], prized not only for their strength but also because it was believed that the mana [spiritual power or energy] of the deceased would render the hook particularly efficacious." The dimensions and shape of certain hooks and their occurrence in the archaeological record

with various fish bones indicate specialization of fishing for a given species. For instance, Hawaiians used particularly large and sturdy hooks
for sizeable bottomfish and pelagic species in the benthic and pelagic
zones (Kirch 1985: 206–208).

Isabella Aiona Abbott, a famous ethnobotanist and first Native
Hawaiian woman to receive a doctoral degree in science (UC-Berkeley
1950) wrote extensively about traditional use of plants and seaweeds in
the Hawaiian Islands. Interviewed in 1999 as part of an investigation of
traditional small boat fishing in the islands (Glazier 2002), Dr. Abbott
revealed a deep knowledge of fishing and fishing gear used in old
Hawai'i. Her work with plant materials and review of historical sources
relating to use of native plants make clear that olonā (*Touchardia lati-
folia*) was the most highly functional and favored material for fishing
line, and that the relationship between aho makers and fishermen was
an essential social aspect of the ahupua'a system (Abbott 1992, 1999).

Goto (1986: 155) reviews Kahā'ulelio's description of the ancient
aku fishery. Nets were sometimes used for aku but trolling with lures
attached to handlines was favored. Poles were used in some fisheries,
but in all cases the lines were skillfully retrieved by hand, with pressure
kept on the line to ensure the fish remained on the hook. While various
other pelagic fish species were pursued and consumed on occasion, the
relative lack of their remains in the archaeological record, and the predominance of aku bones, suggest that aku was a principal and therefore
commonly savored target. Goto (1986) uses elements of the description provided by Kahā'ulelio (1902) to discuss the pursuit of aku by
Hawaiian fishermen of old, a process that involved certain conditions
and indicators:

Kahā'ulelio (1902) mentioned that fishermen would set out to catch
aku only when the sea was calm and smooth enough to see schools
of nehu [Hawaiian anchovy; *Stolephorus purpureus*] and other fishes.
These small fishes are eaten by aku and therefore they are the indicators
of the presence of a school of aku around them. The location of bird
flocks was also a good indicator of aku...In general, the leeward side[s]
of each island offers good conditions for trolling in the pelagic zone
because of [predominant) wind patterns [which render the ocean sur-
face relatively smooth]. (p. 155)

Aku was eaten fresh and/or preserved with salt. The fish was an important trade item and an important food for the aliʻi (Goto 1986: 117–126). Aku provides 129 food calories per edible part, totaling 839 calories per kilo of gross weight. At 25.8 grams protein per 100 edible grams of meat, aku generated the highest percentage of fish protein then available to Hawaiians. As such, the fish provides more protein than wild boar, the meat of which yields roughly 17.5 grams of protein per 100 edible grams. Aku is also an excellent source of vitamin E, with its eyes and organs rich in minerals—there was a long custom of consuming these (Titcomb and Pukui 1951: 29). Seafood, and aku especially, did not merely supplement the Hawaiian diet but was rather an essential component, balanced by root crops, fruits, and other foods. The dietary importance of aku and its seasonal abundance may partly explain its apparent religious significance for early Hawaiians (Valeri 1985: 79).

The palu ʻahi form of fishing has a long history of use in Hawaiʻi, likely beginning long before contact with Europeans. Maly and Maly (2003) describe the basic mechanics of palu fishing methods, based on readings of various historical accounts, including those of Beckley (1883: 9), Kahāʻulelio (1902), and others:

> Fishermen had many customs and devices. The lihi was one kind of hook; another was a baited hook; octopus palu or chum was the device used by some fishermen; released hoʻoholo (live fish) was the bait of others. A hook baited with flesh (paʻiʻo) was another bait. One kind of palu was handfuls of whole fish—ʻōpelu or akule or puhi kiʻi perhaps—pounded until soft, and wrapped in coconut cloth, aʻa niu, with a stone inside. This was let down to the bottom of the koʻa, and then shaken until the stone rolled out and the palu scattered. Those who understood the properties (mana) of good bait would come to shore with a good catch... (Volume I, p. 41)

The modern techniques are similarly straightforward. While few data are sources available to help describe the extent of use of such methods over time, the work of Glazier et al. (2009: 28–31) indicates that the palu ʻahi method was consistently used in the twentieth century by a relatively small number of Native Hawaiians who fished over various koʻa in remote areas along the Big Island, Kauaʻi, and other islands. The method is presently being used by a broader subset of Native Hawaiian and other local residents who recognize its simplicity and efficiency (Fig. 3.3).

Hawaiians developed distinct kinds of palu and approaches for attracting different species. For instance, vegetable matter such as taro and pumpkin were traditionally used to attract and train 'ōpelu at 'ōpelu ko'a along the Kona side of Hawai'i Island (Abbott 1999; Glazier 2007: 91). Palu made from 'ōpelu was often used above 'ahi ko'a. Appropriate formulation and use of palu has long been of concern to culturally savvy Native Hawaiian fishermen (Beckley 1883: 6).

Efficient hooks and strong cordage obviously were (and remain) critical to fishermen in Hawai'i. But the fishermen could not reach the fishing grounds outside the reef or in the open ocean without a study vessel. Until, and even into the twentieth century, this was for many Native Hawaiians the wa'a (outrigger canoe), typically made primarily from the strong, durable, and buoyant koa (*Acacia koa*). Dr. Robert Hommon, former lead U.S. Department of the Interior archaeologist at the Hawai'i Volcanoes National Park in Hilo, asserts that most canoes were used for benthic and pelagic fishing and estimates that as many as

Fig. 3.3 Native Hawaiian fishermen prepare for an offshore trip at ancient launch site in Halawa Valley, Moloka'i

12,000 canoes were active in the Hawaiian Islands at the peak of their use (Hommon 2000).

Kahāʻulelio (2006: 25) relates that different types of canoes were used for different kinds of fishing. Those used for aku were called hoʻomo. Corney (1896) observed canoe-based aku fishing in the late nineteenth century and described the process as follows.

> A canoe that pulls seven paddles goes to sea with two good fishermen (besides the paddlers), each with a stout bamboo, about 20 feet long, and a strong line made from the oorana [olonā]…the line is about three-quarters of the length of the pole and has a pearl hook made fast to it. The canoe is then paddled very swiftly with the hooks touching the surface of the water, one at each side, the fisherman holding the rod steady against their thigh, and the lower end resting on the bottom of the canoe; they steady the pole with one hand, and, with the other keep throwing water on the hook [ostensibly stirring up the palu and imitating the effect of surface feeding], and when their prey gets hooked, by lifting the pole upright the fish swings in, and is caught under the left arm and secured. In this manner they will take 40 to 50 in the course of a few hours. (p. 209)

Sail power was used when possible, but when the wind was calm the power of paddlers was needed. Sometimes special double canoes with malau (baitwells) would be used, and nehu or ʻiao (Hawaiian silverside; *Atherinomorus insularum*) would be added to the water to stimulate feeding behavior, a process that can be observed on certain vessels around the islands today (Glazier et al. 2013: 357). Kahāʻulelio notes that when malau canoes were used, single canoes could not come close or would suffer the penalty of gear confiscation; he also states that his father and many others had given up on malau fishing around 1848, since it involved too much work (Kahāʻulelio 1902: 39). This may indicate that the social system that supported communal fishing operations in that region in years past was beginning to falter.

Kamakau (1976) provides a similar description of aku fishing but also discusses the use of lures, trolled behind canoes in the manner of modern tolling vessels. He notes that this kind of fishing was under-taken by chiefs and rulers (who, in keeping with their status, may have fished more than paddled), and offers a detailed description of the red cowry aku trolling lure:

> At the base of the pā there is a ridge, and through this was a hole drilled as a foundation for the cord of the snood, ka'ā. The cord ran from the hole to the edge of the hook that was fastened to the tip of the shank lure, pā. The hook was made of human or dog bone, filed smooth and curved nicely. Pig bristles crossed the base of the hook where it joined onto the tip of the shank so that the hook would not fall over. The bristles ruf-fled the water behind the lure as those on the canoe paddled in unison, and the aku mistook the lure for an 'iao or other small fish and crowded around to seize the pā hī aku. (p. 75)

'Ahi and other large fish such as a'u (billfish, including black, blue and other marlins, were also pursued from canoes. The challenges of pur-suing marlin from paddled canoes were especially formidable since the fish can occasionally far exceed 1000 pounds (though large quantities of meat can obviously be obtained by landing such creatures, today the meat of such large fish is typically smoked). The large pelagics, includ-ing various billfish species that are not allowed to be taken by fleets around the U.S. continent, remain a particularly important food source among contemporary Native Hawaiian and local fishing families around the islands (Fig. 3.4).

Social Dimensions of Lawa'ia

Abbott (1999) asserts that because relatively few persons lived in any given ahupua'a in very ancient times, and because other food resources were plentiful, the need to pursue deep sea fish species was probably not as critical as during later eras. But as the ali'i set themselves apart from the rest of the population, surplus economies developed, and open ocean fishes became increasingly important items of trade. Goto (1986: 451) states that ali'i probably began to require increased har-vest of pelagic species as part of a developing socioeconomic system in

Fig. 3.4 Large aʻu (marlin) to be shared among ʻohana in Waiʻanae, Oʻahu, mid-2000s

which such fish were desired items for tribute. Yet, as made apparent by Titcomb and Pukui (1951), while fish were distributed in relation to a sociopolitical hierarchy, this incorporated concern for community well-being:

After fish were offered or set aside for offering by giving them to the priest, the best fish of the catch were set aside for the chief in an amount to provide generously for his personal needs and those of his numerous household. Then the various *kāhuna* (recognized experts in branches of learning), next the *konohiki* (chiefs agent and overseer), and finally the people received their share. Division was made according to need, rather than as reward or payment for share in the work of fishing. Thus all were cared for. Anyone assisting in any way had a right to a share. Anyone who came up to the pile of fish and took some, if it were only a child, was not deprived of what he took, even if he had no right to it. It was thought displeasing to the gods to demand the return of fish taken without the right. What Hawaiians thought sometimes about this inevitable sharing of a hard won catch may be known from the following lines from the legend of Niho'o leki…"The current is flowing towards Maka'ena, where swarm the aku, where the giving would be a pleasure, when the worthless could have a share, when the hungry of the uplands of Waiahulu could have a share". (p. 8)

As the ahupua'a became less integrated in the mid-nineteenth century, the effort and products of fishing and shoreline gathering would gradually have become more specific to the 'ohana than to ahupua'a societies as a whole. This is not to say that inter-familial social and economic relations ceased. To the contrary, these would have become increasingly important. Rather, the ahupua'a as a distinctly functioning sociopolitical system—in which tribute was given to the ali'i through a hierarchy of relations—would have been diminished. But sharing, bartering, customary exchange, and other forms of reciprocity-based transfer of seafood and other resources continued to sustain the maka'āinana into the twentieth century (Andrade 2008; McGregor 2007). Such social and economic relations, in conjunction with involvement in the contemporary market economy, characterize life in many Native Hawaiian and other local 'ohana and communities to this day (Allen 2013; Glazier et al. 2013; Glazier and Kittinger 2012; Vaughan and Ayers 2016; Vaughan and Vitousek 2013). This is also true of other U.S. Pacific Islands (Severance et al. 2013).

As is the case among many island fishermen today, canoe fishing in ancient Hawai'i was an organized venture involving the participation of experts whose knowledge of the ocean and its resources maximize the likelihood of success. Oliver (2002: 89) discusses such specialization as a common aspect of Polynesian societies in general. In the case of old Hawai'i, the lawai'a haku were differentiated from lawai'a 'ili'ili or assistant fishermen (Kamakau 1976: 4–5), indicative of a hierarchy of experience, skill, knowledge, and related status. This, too, is reflected in modern fishing activities across the islands (and elsewhere)—youth are often introduced at an early age. Some kids continue to participate over time and gradually gain extensive knowledge, status, and satisfaction from so doing,

Hawaiian society was task-oriented and specific duties were assigned to males or to females. Although wahine (women) were not generally part of the actual act of benthic or pelagic fishing in ancient times, they did participate extensively in critical preparatory activities, in the processing of fish, and as mentioned above, in various shoreline gathering activities (Ka'ai'ai 2018). Abbott (1992: 84–85) states that women fished with traps, baskets, and other gear along the nearshore reef systems of the islands, but apparently boat fishing and the gear used on boats was the exclusive domain of men. Titcomb and Pukui (1951: 4) state that "every day saw many people, women in the majority, out on the reefs for hours, searching, collecting all that was edible and desirable."

With respect to fishing on the open sea, the expert fisherman was accorded high status in ancient Hawaiian society. He knew best how to judge weather and sea states and signs indicating the presence of fish. He was well-versed in navigational skills and knew how to manage the interaction of fellow fishermen. Such knowledge and skills continued to be valued in Hawai'i fishing ports and communities in this modern era (Glazier 2002, 2007; Glazier et al. 2013). As discussed by Kamakau (1976: 76), and as it remains today, fishing knowledge, including the location of productive fishing spots, was invaluable in its capacity to feed the 'ohana and larger community. It was and is therefore transmitted judiciously, if at all:

> Ka poe kahiko regarded their secret fishing grounds, koa huna, as "calabashes and meat dishes" (he umeke a he ipu kai) and as "grandparents" (kupuna kane a he kupuna wahine) [sources of provisions] and could be robbed and beaten before they would reveal their location. They pointed out their secret fishing grounds only to their own children. (p. 76)

At sea, communication between fishing canoes enabled a coordinated operation. Under the direction of the lawai'a haku, fishermen of old signaled to each other by waving their arms or bamboo poles (Titcomb and Pukui 1951: 5). Beckley (1883) provides an indication of the depth of skill and knowledge possessed by fishermen of old Hawai'i. Knowledge of habitat and fish behavior, ability to navigate, appropriate equipment, and skill were and continue to be essential to successful fishing in the deep ocean waters surrounding the islands:

> For deep-sea fishing the hook and line are used without rods, and fishermen sometimes used [hand]lines over 100 fathoms in length. Every rocky protuberance, from the bottom of the sea for miles out in the water surrounding the islands, was well-known to the ancient fishermen, and so were the different kinds of rock where [bottom-feeding] fish likely to be met...The ordinary habitat of every known species of Hawaiian fish is also well known to them. They often went fishing so far out from land as to be entirely out of sight of the low lands and mountain slopes and took their bearing for the purposes of ascertaining the rock which was the habitat of particular fish they were after from the position of the different mountain peaks. (p. 10)

Although Geographic Positioning System (GPS) technology is typically used today to determine one's location on the ocean and one's location relative to known fishing locations, many local fishermen remain adept at triangulating by landmarks. Similarly, although fishermen of the present day use cell phones and VHS radio to communicate essential information, the core content of the typical discussion that was formerly signaled—birds, bait, fish, approach, and position—remains the same.

Spiritual and Religious Dimensions of Lawa'ia

Titcomb and Pukui (1951: 41) assert that the lawai'a hāku not only possessed functional knowledge of the sea and fishing but were also capable of interpreting dreams and omens. Such persons maintained a good relationship with the gods, including one's 'aumakua (family god, or personal guardian, assuming the form of an animal). Abbott (1999) relates understanding from her kūpuna that ancient Hawaiians offered much prayer and conciliation to Kāne and other gods of the sea prior to the fishing trip.

Fishermen of old observed strict kapu and kānāwai before entering the ocean, such as avoidance of certain foods, conversations, and social contacts (Titcomb and Pukui 1951: 39–41). Fishing thereby involved both a spiritual dimension and ongoing communal involvement, and a successful trip led to an offering of appreciation to Akua (God, the gods), followed by proper distribution of the fish among those involved. Similar behaviors, including ritualistic preparation before fishing, special behaviors during the trip, offering of the first catch of a given fishing season, and sharing the catch in familial and community settings can be observed of certain small boat fishermen in Hawai'i today (Glazier 2007; Glazier and Kittinger 2012; Glazier et al. 2013). The words of Kamakau (1976) indicate that fishing of old involved actions that were both spiritual and economic in nature:

> He (the master fisherman) cast down the fish for the male 'aumakua (family god) and for the female 'aumakua, and then returned to give the fish to the canoe men, to those who had done the chumming, and to those who had done the actual fishing. A portion went to the owner of the fine mesh nets, nae puhi, that had been used to catch bait and to those who had driven the bait fish into the nets. The rest was for the head fisherman or for the land holder, if it had been the land holder's expedition. (p. 74)

Goto (1986: 437) reports that, because aku bones are sometimes found in association with inland religious sites, the fish were probably valued as ritualized offerings to the gods. He also suggests that the seasonal nature of aku operations and associated rituals and patterns of consumption contributed to the ordering of ancient Hawaiian society. A general review of Kirch (1985), Kirch et al. (2015), Goto (1986), and other authors makes clear that much effort was involved in the construction of fishing heiau on each of the main islands. Notably, Native Hawaiians continue to make offerings at heiau in association with important cultural activities, including fishing (Glazier 2007: 16). Maly and Maly (2003 II: 418) also provide instructive discussion

of mōhai, or sacrificial offerings, and hoʻokupu (tribute) wherein portions of the first ʻahi of the season are sent back to the god of sea and
meat is shared with the community from which the fishermen hail
(Fig. 3.5).

Successful negotiation of small-boat deep-sea fishing around the
Hawaiian Islands in the winter season, when massive swells propagate
southward from intense storms in the Gulf of Alaska, requires intense
concentration, fortitude, and luck. Success in such cases can be defined
simply in terms of making it back to port. But the need to land fish
can inspire a trip irrespective of ocean conditions and success can satisfy
critical dietary and economic imperatives. This has been true for mille-
nia and continues to be the case in the modern era—as made clear in
the following summary of a Christmas Day fishing trip from the port
town of Haleʻiwa on Oʻahu's North Shore in 1993 (Stecyk 1993):

> 4:45 a.m. The harbor is completely silent. In the new-moon darkness
> there's only the steady thunder of giant waves breaking far out on the
> cloudbreak reefs. Faith is our guide as we throttle the *Hoʻolina* past the
> moorings…It's unlikely other boats will venture out today…Fishermen
> must take chances in hard economic times, but a bad catch today would

Fig. 3.5 Heiau at rugged Ka Lae in Kāʻu district, the southernmost point in
Hawaiʻi and the 50 states

certainly dishonor us in the eyes of our friends, lovers, spouses and kids. But a good catch—ah!…Suddenly, each of our six lines is hit! It's an angry spectacle of thrashing iridescence as a school of mahimahi charges the boat in a feeding frenzy. Rapaciously we cast, hook, land, de-hook and re-cast, working against time and exhaustion. Finally, in an instant, the fish are gone. A cursory tally yields 38 large mahi…Minutes later, an immense black marlin erupts in an explosion of defiance, and the game begins in earnest…Recognizing the magnificence of the 500-plus-pound beast [and having already kept all the fish that were needed for the purposes of the trip at hand], we carefully release the fish and watch him speed back down into the blue-black depths. That night we learned that the exact time (11:05) we snagged the marlin, veteran waterman Titus Kinimaka has his femur snapped by a huge wave [while surfing in the Eddie Aikau big-wave contest] at Waimea Bay…Liko reckons that "our" marlin was Titus' uhane kia i, or guardian spirit. Either way, Mele Kalikimaka! [Merry Christmas!] (p. 128)

Kapu Lawaiʻi

Previous sections of this chapter have provided evidence that the abolition of kapu and the affinity of certain aliʻi to new people and new ways of conceiving the universe and its creator during the eighteenth and nineteenth centuries exerted only a superficial effect on the makaʻāinana. Again, there was much social resistance to change, and Hawaiians did not readily leave behind the ways of the past since a core attribute of the culture has always emphasized ka lamakū o ka naʻauao (the torch of wisdom) available through the kūpuna and their experiences.

Based on the need for careful use and management of the resources of land and sea during periods of social and economic upheaval, the basic principles underlying early kapu and kānāwai, and certain of the strictures themselves, continued to be used into the twentieth century and, in various places around the islands into the current century. This is evinced by: (a) an extensive literature that is indicative of persistent use of fishing kapu and kānāwai around the islands over time (e.g., Titcomb and Pukui 1951; Handy et al. 1972; Iversen et al. 1990; Poepoe et al. 2003; Chun 2009b; Jokiel et al. 2011; Maly and Maly

2003); (b) interviews and discussions with living and recently deceased kūpuna who speak extensively of kapu and how marine resources were cared for in the past and should be cared for in the present; and (c) recent and ongoing observation of and participation in fishing-related community life around the islands, which, in the Native Hawaiian context, consistently indicates ongoing use of kapu and other traditional management strategies (e.g., Glazier 2002, 2007; Glazier et al. 2009; Pacific Islands Fisheries Science Center 2018; Vaughan and Ayers 2016; Vaughan and Vitousek 2013). The nature and manner of contemporary use of place-specific kapu and other management strategies was a core subject of the many hundreds of hours of group deliberation that are examined in the following chapter of this book.

A useful prefatory discussion of the historic role of konohiki and kāhuna lawaiʻa in caring for marine resources and thereby ensuring their continued use in a given ʻili, ahupuaʻa, or moku is provided by Jokiel et al. (2011: 3):

> Fishermen were of a special lineage and trained for years as an apprentice. During this time they were taught to observe subtle and major changes in the condition of the marine resources. They were educated in the life cycle, diet, daily and seasonal feeding habits, preferred habitat, and growth conditions. They obtained knowledge of the appropriate season, time of month, time of day, and method for harvesting of the many species of fishes, invertebrates, and seaweeds. Harvest management was not based on quota, but on identifying the specific times and places that fishing could occur so that it would not disrupt the basic habits of important food resources nor deplete fish stocks (Poepoe et al. 2003)…The belief was held that resources were limited and there was a social obligation to exercise self-restraint in resource exploitation. The ancient Hawaiians viewed themselves as an integral part of nature (Kawaharada 2006; Maly and Maly 2002, 2003; Kirch 1985; Valerio 1985).

The discussion (Jokiel et al. 2011: 3) goes on to note the importance and evolving nature of traditional ecological knowledge and its application in terms of managing a base of precious food resources:

> Hawaiians possessed a complex understanding of the life histories of fishes. Perceptive observations led to a keen familiarity of physical (e.g.,

weather patterns, currents, tides, wind, waves), biological (e.g., spawning seasons, recruitment, and growth), and ecological (e.g., foraging patterns, behavior, and habitat) factors that influence fisheries. In these areas the traditional knowledge of Hawaiian fishermen may have surpassed what is known by modern marine biologists (Gosline and Brock 1960; Lowe 2004). Knowledgeable *kūpuna* also consulted with *poʻo lawaiʻa* (master fishermen) who had intimate awareness of the status of various populations of organisms. When populations declined to low levels, a *kapu* (forbidden practice) was placed on extraction to allow the resource to recover (Maly and Maly 2002, 2003). (p. 3)

The words of Titcomb and Pukui (1951) similarly reflect the importance of rules and discipline in governing use of natural resources in the ahupuaʻa of old Hawaiʻi. Of note in the text is mention not only of external enforcement of appropriate behavior, but also one's internal guidance that stems directly from Hawaiian culture and associated beliefs. Territorial and, later, state and federal authorities and regulations, gradually superseded the authority of the konohiki, at least at a superficial level. External authority continues to frustrate the sensibilities of Kānaka Maoli whose knowledge and wisdom derive from long years of observation, trial and error, and inter-generational knowledge. In certain rural ahupuaʻa around the mokupuni (islands) today, such practitioners continue to exert a strong cultural influence, and use of natural resources is essentially locally governed. A case in point is the traditional custom of asking local leaders for permission to fish, gather, hunt, or otherwise use natural resources in and offshore from a given ahupuaʻa. This is still enforced in some areas, as violators may soon discover.

Basic incongruites between traditional and modern resource management strategies are further noted in the following chapter. Here Titcomb and Pukui (1951) hearken to a time when the traditional authority of the gods was believed to manifest in human agents, comprising an interconnected system of divine and human oversight of precious natural reosurces. Enforcement measures of those days now appear draconian, but it must be remembered that survival of society itself was at stake:

Tabus were an instrument in the conservation programme. The political power was concentrated in the upper class, the chiefs, and the laws of the land and of the sea were their edicts. The penalties for breaking tabus were heavy, often the death penalty for what seems to us a trifling fault. This held the people in a strict discipline. Besides tabus, the relationship with the gods was a powerful determinant of action. The lesser gods that each person had, personal gods, as well as the greater gods whose power was universal, were ever present. Their will was interpreted through the priests but understood well by the people too. To conserve resources was a custom rigidly adhered to. It was the will of the chiefs, and also the will of the gods, and it was obviously wise. When a man broke this law he expected punishment from the chief's agent (*konohiki*), if his act was detected, but punishment from the gods certainly, for no knowledge was hiddden from their perception. Man appealed to his gods for good luck, but the gods expected man to do his share in making it possible. (p. 13)

Persistence Among Native Hawaiian Fishermen

Hawaiians continued to fish during and after the mahele, and as the new century approached, new materials were sometimes adapted for use in the present. Scobie (1949: 289) states that "many of the changes were substitutions of materials in old forms...spears were tipped with iron, metal hooks were fastened into the cowrie shell by molten lead, floats might be of cork and sinkers of iron or lead...new methods came with new fishermen." In some areas, fishhooks were made from metal wire and nails. But notably, the author (ibid.) indicates a tendency toward self-reliance among Native Hawaiians, stating that "[many] still preferred the fishhooks they made themselves rather than those that could be bought."

Old methods and new materials were sufficient to contribute significant amounts of seafood to populations of Chinese, Japanese, Filipino, and Portuguese plantation workers and others living in Hawai'i during the Plantation era. Cobb (1905: 752) notes that the first regular market for the sale of fish was established in Honolulu in 1851, and that Native Hawaiian fishing operations, including fishponds, were the principal source of seafood. U.S. Commission of Fish and Fisheries statistics (Territorial section) reports that the 2345 persons who sold fish in Hawai'i in 1901, included 1571 Hawaiians, 485 Japanese immigrants, and 238 Chinese immigrants. These figures shifted rapidly in subsequent decades, as market demands shifted and some immigrant plantation workers, especially those from Japan, began to compete with Native Hawaiian canoe fishermen in the commercial marketplace. Some such persons had come to Hawai'i with the specific intent of fishing on a commercial basis and a few were supported with capital from Hawai'i-based Japanese families in order to do so (Schug 2001: 20; Ogawa 2015).

Subsistence-oriented fishing and the availability of commercially landed seafood helped enable the survival of a severely diminished population of indigenous Hawaiians at the start of the twentieth century, then enumerated at 29,790 persons, with another 7857 persons reporting part-Hawaiian ancestry. The total population of the islands at the turn of the century was enumerated at 154,601 persons (U.S. Department of Commerce, Bureau of the Census 1913).

Given centuries of profound social change, there was little capital available in Hawaiian communities during the Plantation era, and there are clear indications of basic economic difficulties in fishing-oriented households. For instance, Abbott recalls of her childhood in the 1920s that unless olonā had been saved from decades past, Hawaiians would use other household materials for fishing line. Abbott's uncle, for example, used lines made from coconut fiber or butcher string (Abbott 1999).

Gaps in the historical record do not allow for a clear account of Native Hawaiian fisheries in the years following the rise of the Japanese commercial aku fleet and growing predominance of the Japanese sampan-style fishing vessel in Hawai'i during the 1920s and 1930s. Authors living through that period tended to focus on historical aspects of the region's fisheries rather than documenting the nature of fishing during their own lifetimes. Undoubtedly there was continuity in use of outrigger canoes, old-style line and lures, and fishing methods. But data are sparse. Landings data provide no clear clues as to the scope and scale of subsistence fisheries around the islands since only commercial landings data were collected by government agencies between 1900 and 1986.

Fig. 3.6 Winter season spear fishing from traditional three-board waʻa, South Kona coastline, mid-1990s

There is indication that the status of certain fish populations had declined by the turn of the twentieth century. Titcomb and Pukui (1951: 17) suggest that this was related in part to diminishing control of fishing activity by the konihiki. Although most Native Hawaiian persons who would be able to speak empirically of the pre-World War II era have now passed on, Maly and Maly's (2003) work with kāhuna lawaʻia and elders around the islands provides evidence that Native Hawaiian fishing traditions and ecological knowledge continued to evolve into the mid-twentieth century, irrespective of socioeconomic effects stemming from the hegemonic tendencies and capitalist practices of the haole. Apparently, many Hawaiians remained dedicated to traditional approaches to fishing. For example, as required of the old fishing methods, many Hawaiian fishermen preferred the relative silence of the paddled canoe to the first vertical crankshaft (outboard) motors (Abbott 1999) that were mass-produced by Ole Evinrude and available for use on small boats in Hawaiʻi by about 1910.

Modern materials eventually replaced the old even in the more remote areas around the islands, where traditions tended to linger in what McGregor (2007: 4) refers to as kipuka (pockets of traditional culture). For instance, now deceased participants in Glazier's (2002) work on social aspects of Hawaiʻi's small boat fisheries stated that, by about 1940, the waʻa pā (three-board canoe) had largely replaced the traditional koa dugout in most villages along the Kona Coast of Hawaiʻi Island. The three-board canoe resembles the traditional dugout outriggers, even in the manner of its lashings. But it also utilized steel and, later, aluminum pipe for ʻiako (booms), metal screws and plates, and outboard engines. The craft were commonly used at Hoʻokena and Puʻuhonua O Hōnaunau on Hawaiʻi Island as late as the late 1990s and continue to be used in certain West Hawaiʻi villages today (Fig. 3.6).

Local Japanese fishermen were denied access to the ocean during World War II (Yuen 1979: 7), and many boats were confiscated. Schug (2001: 29) reports that at least six such fishermen were killed at sea by U.S. naval forces who believed they were conspiring with the enemy during Japan's attack on Pearl Harbor. Many sampans and other fishing vessels were reinstated at war's end. Numerous larger sampans were used for commercial fishing, typically for aku. Some of the smaller boats were used by local Japanese ʻohana to fish for food with relatively little operating cost. As discussed by Heeny Yuen, respected kupuna and employee of NOAA Fisheries and its predecessor agency from the late 1940s through 1991, Native Hawaiians continued to use canoes and other small craft to fish commercially and for food during and after Hawaiʻi was deemed one of the United States on August 21, 1959 (Yuen 2016). According to the late Walter Paulo, highly respected kahuna lawaʻia from Hawaiʻi Island, Native Hawaiian fishermen established the first longline fishery at Cross Seamount during the 1940s (Paulo 2002).

The small-boat fleets grew dramatically across the islands in conjunction with an improving post-war economy and improvements in hull and engine technologies. Introduction of the Johnson 55 horsepower outboard motor in 1968 revolutionized small-boat fishing with increased power, efficiency, and especially reliability at sea. Fiberglass hulls and marine communication technology also progressed and became more affordable during the 1960s and 1970s. This was also a time of economic expansion in the islands. Construction activity peaked at 8.2% of the Gross State Product in 1970 (State of Hawai'i, Department of Business, Economic Development and Tourism 2000), with most new construction occurring on O'ahu. The sugar and pineapple industries still employed many workers on most of the islands at that time. Relatively safe and efficient small fishing vessels thus became accessible during a time when many in Hawai'i's working class could afford to buy and operate them.

Fishing from the shoreline became increasingly popular in the post-World War II era. The activity continues to be very popular across the islands, as do trolling for pelagics and static fishing for various benthic species. Gear, vessels, engines, and sensing and communications continue to evolve toward greater efficiency and cost. Today, many Hawai'i residents cannot afford such investment, and thus more affordable shoreline and reef-based fishing and diving activities are undertaken. Given the financial and logistical challenges of owning, maintaining, and trailering or mooring a seaworthy boat in the islands, many shoreline fishermen report satisfaction with the simplicity and relatively inexpensive nature of fishing from the shore, though many residing in populous island areas also report dissatisfaction with catch rates. Kayak-based fishing is increasingly popular, enabling ocean access without great investment. Notably, shoreline fishing is occasionally undertaken using traditional strategies, as demonstrated by fishermen Ben Hauanio and Aku Hauanio, who meet the challenges of fishing along the very rugged Puna coastline of Hawai'i Island using age-old pole and line methods to capture ulua (Langlas 2003). Irrespective of the many challenges associated with fishing, including the challenges associated with a growing island population and a finite base of natural resources, the activity provides a source of food and recreation for many Native Hawaiians and residents of all ethnicities in the twenty-first century.

The Twenty-First Century

Relatively few island or coastal residents around the world today are in a position to undertake subsistence-oriented natural resource harvesting activities on a full-time or exclusive basis. This holds true in the Hawaiian Islands, where fishing, hunting, and other ocean- and mountain-based food gathering activities are parts of an overall suite of activities and sources that contribute to the economic viability of many Native Hawaiian and local households. These include part-time and full-time jobs (often in the tourism-related service and construction industries) small-scale agriculture, returns on investment, limited subsidies, customary exchange, and intra- and interfamilial sharing of labor, food, and fiscal resources. Commercial fishing and income-bearing sub-rosa activities are also important in economic terms in certain family and community settings around the islands.

By combining many sources of food and income, local residents are able to take part in historically important and culturally meaningful activities that bear tangible benefits for the family unit and, collectively, for a larger community of involved residents. In many areas, this simultaneously involves maintenance of local and traditional ecological knowledge and strategies for sustaining populations of living marine resources for future use.

Observation in Hawai'i community settings makes clear that most fishermen take to the shoreline and open-ocean to capture food for consumption in nuclear and extended family settings. The activities are often enjoyable to the participants, and, in this sense, it may be said that a recreational experience is involved. But fullest enjoyment tends to be observed or reported when a successful harvest occurs and there is food from the sea (or mountains) to eat, to share, and to instigate or satisfy social interaction. In some cases, these experiences are said by the participants to involve a spiritual connection with the natural environment, with one's family, and with creative powers greater than the individual.

Although sociodemographic and economic changes continue to alter contemporary life in Hawaiian 'ohana and communities, kūpuna continually assert the importance of: ensuring the continuation of local traditions, ecological knowledge, and prudent use of resources by residents and non-residents alike; pursuing and consuming seafood in family and community settings; and deriving dietary, social, and household-economic benefits from successful fishing, gathering, and hunting activities. Notably, the population of Native Hawaiians and part-Native Hawaiians around the state now approaches 300,000 persons as the new decade nears (Goo 2015), with the overall population estimated at 1.4 million persons. As discussed in the following chapter, the interest of many Native Hawaiians in maintaining traditional aspects of family and community life is deeply held even as populations grow and marine and coastal environs are increasingly pressured. Neglecting such interests in the natural resource management process could potentially serve to constrain customs and traditions that were and remain fundamental to the development a Polynesian society that is now growing and exerting its capacity and right to manage natural resources across the archipelago (Fig. 3.7).

Fig. 3.7 Esteemed kahuna lawaʻia setting up for a night of handlining along Hawaiʻi Island

References

Abbott, I. A. (1992). *Lā'au Hawai'i—Traditional Hawaiian Uses of Plants*. Honolulu: Bishop Museum Press.

Abbott, I. A. (1999, January). Personal Communication. Department of Botany, University of Hawai'i at Mānoa,Honolulu.

'Aha Pūnana Leo. (2018). *A Timeline of Revitalization. E Ola Ka 'Ōlelo Hawai'i—The Hawaiian Language Shall Live*. Hilo. Available at http://www.ahapunanaleo.org/index.php?/about/a_timeline_of_revitalization/.

Allen, S. (2013). Carving a Niche of Cutting a Broad Swath: Subsistence Fishing in the Western Pacific. *Pacific Science, 67*(3), 477–488. In Special Issue of *Pacific Science*—Human Dimensions of Small-Scale and Traditional Fisheries in the Asia-Pacific Region (J. Kittinger & E. W. Glazier, Eds.).

Andrade, C. (2008). *Hā'ena: Through the Eyes of the Ancestors*. A Latitude 20 Book. Honolulu: University of Hawaii Press.

Beckley, E. M. (1883). *Hawaiian Fisheries and Methods of Fishing with an Account of the Fishing Implements Used by the Natives of the Hawaiian Islands*. Honolulu: Advertiser Steam Print.

Blaisdell, K. (1996, April 28). Historical and Philosophical Aspects of Lapa'au—Traditional Kanaka Maoli Healing Practices. *In Motion Magazine*. Available at www.inmotionmagazine.com/kekuni3.html.

Bushnell, O. A. (1986). Foreword to *Treasury of Hawaiian Words: In One Hundred and One Categories* (I I. W. Kent, Ed.). Honolulu: University of Hawaiian Press.

Bushnell, O. A. (1993). *The Gifts of Civilization—Germs and Genocide in Hawaii*. Honolulu: University of Hawai'i Press.

Campbell, A. (1819). *A Voyage Around the World*. New York: Broderick and Ritter.

Chan, D. (1994, August). *Kāhuna Lā'au Lapa'au: Issues and Concerns Involved in Potential Licensure* (Unpublished paper). University of Hawaii School of Public Health. On file with Papa Ola Lokahi. University of Hawai'i at Mānoa, Honolulu.

Chan, H. L., & Pan, M. (2017). *Economic and Social Characteristics of the Hawaii Small Boat Fishery 2014*. U.S. Department of Commerce. NOAA Technical Memorandum NOAA-TM-NMFS-PIFSC-63. 107 pp. https://doi.org/10.7289/V5/TM-PIFSC-63.

Chun, M. N. (1989). *Ka Mo'olelo Laikini Lā'au Lapa'au: The History of Licensing Traditional Native Practitioners*. Honolulu: Hawai'i State Department of Health. Health and Education Branch.

Chun, M. N. (2009a). *It Might Do Good: The Licensing of Medicinal Kāhuna.* Honolulu: First Peoples' Productions.

Chun, M. N. (2009b). *Ho'onohonohono—Traditional Ways of Cultural Management.* Ka Wana Series. Curriculum Research and Development Group, University of Hawi'i at Mānoa, Honolulu.

Churchill, W., & Venue, S. H. (2004). *Islands in Captivity—The International Tribunal on the Rights of Indigenous Hawaiians.* Cambridge, MA: Southend Press.

Clark, J. R. K. (2011). *Hawaiian Surfing: Traditions from the Past.* Honolulu: University of Hawaii Press.

Cobb, J. N. (1905). *The Commercial Fisheries of the Hawaiian Islands: III* (Reprinted in the U.S. Fish Commission, Bulletin No. 23).

Connors, R. H. (2009). *Gender, Status and Shellfish in Precontact Hawaii.* Thesis completed in partial fulfillment of the requirements for the degree Master of Arts in anthropology. San Jose State University, San Jose.

Corney, P. (1896). *Voyages in the Northern Pacific, 1813–1818.* Honolulu: Thomas G. Thrum.

Damon, S. C. (1869). *Puritan Missions in the Pacific: A Discourse Delivered at Honolulu, (S. I.,) on the Anniversary of the Hawaiian Evangelical Association.* Sabbath Evening, June 1866 (First American ed., Rev. H. Bingham, Ed.). Printed for J. Hunnewell by Tuttle, Morehouse & Taylor. New Haven, CT.

Department of Hawaiian Home Lands. (2018). *Summary of Hawaiian Homes Commission Act as Amended.* Available at http://www.capitol.hawaii.gov/hrscurrent/vol01_ch0001-0042f/06-Hhca/HHCA_.htm.

Donlin, A. L. (2010). When All the Kāhuna Are Gone: Evaluating Hawa'ii's Traditional Hawaiian Healers Law. *Asian-Pacific Law and Policy Journal, 12*(1), 211–248.

Geslani, C., Loke, M., Takenaka, B., & Leung, P. (2012). *Hawaii's Seafood Consumption and Its Supply Sources.* SOEST Publication 12–01, JIMAR Contribution 12–379. Pelagic Fisheries Research Program, University of Hawai'i at Mānoa, Honolulu.

Glazier, E. W. (2002). *A Sociological Analysis of Fishing Hawaiian-Style.* Dissertation submitted to the Graduate Division of the University of Hawai'i at Mānoa in partial fulfillment of the requirements for the Doctorate in Sociology, Honolulu.

Glazier, E. W. (2007). *Hawaiian Fishermen.* Belmont, CA: Wadsworth-Cengage Publishers.

Glazier, E. W. (Ed.). (2011). *Ecosystem-Based Fisheries Management in the Western Pacific.* Hoboken, NJ: Wiley-Blackwell.

Glazier, E. W., & Kittinger, J. (2012). *Fishing, Seafood, and Community Research in the Main Hawaiian Islands: A Case Study of Hanalei Bay, Kauaʻi* (Final Technical Report). Prepared for the State of Hawaiʻi, Division of Aquatic Resources, Honolulu.

Glazier, E. W., Carothers, C., Milne, N., & Iwamoto, M. (2013). Seafood and society on Oʻahu in the Main Hawaiian Islands. *Pacific Science, 67*(3). In Special Issue of *Pacific Science*—Human Dimensions of Small-Scale and Traditional Fisheries in the Asia-Pacific Region (J. Kittinger & E. W. Glazier, Eds.).

Glazier, E. W., Shackeroff, J., & Carothers, C. (2009). *A Report on Historic and Contemporary Patterns of Change in Hawaiʻi-Based Pelagic Handline Fishing Operations.* SOEST Publication 09–01. Pelagic Fisheries Research Program, University of Hawaii at Mānoa, Honolulu.

Goo, S. K. (2015, April). *After 200 years, Native Hawaiians Make a Comeback.* Facttank: Pew Research Center. Available at https://www.pewresearch.org/fact-tank/2015/04/06/native-hawaiian-population/.

Gosline, W. A., & Brock, V. E. (1960). *Handbook of Hawaiian Fishes.* Honolulu: University of Hawaiʻi Press.

Goto, A. (1986). *Prehistoric Ecology and Economy of Fishing in Hawaii: An Ethnoarchaeological Approach.* Dissertation submitted in partial fulfillment of the requirements for the doctoral degree in anthropology at the University of Hawaiʻi at Mānoa, Honolulu.

Groube, L. M. (1971). Tonga, Lapita Pottery, and Polynesian Origins. *Journal of the Polynesian Society, 80,* 278–316.

Handy, E. S. C., Handy, E. G., & Pukui, M. K. (1972). *Native Planters in Old Hawaii—Their Life, Lore, and Environment.* Bernice P. Bishop Museum Bulletin 233. Honolulu: Bishop Museum Press.

Hasager, U., & Kelly, M. (2001). Public Policy of Land and Homesteading in Hawaiʻi. *Social Process in Hawaiʻi, 20,* 1–31. Available at http://www2.hawaii.edu/~aoude/ES350/SPIH_vol40/10HasagerKelly2001.pdf.

Holmes, T. (1993). Provisions for Polynesian Voyages. In *The Hawaiian Canoe* (2nd ed.). Hanalei, Kauaʻi, Hawaiʻi: Editions Unlimited Publishers.

Hommon, R. J. (1976). *The Formation of Primitive States in Pre-contact Hawaii.* Dissertation submitted in partial fulfillment of the requirements for the Doctorate in Anthropology at the University of Arizona, Tucson.

Hommon, R. J. (2000, May). Personal Communication. U.S. Department of the Interior. Hawaiʻi Island: National Park Service.

Hong, C. (2013). *The Power of the Hula: A Performance Text for Appropriating Identity Among First Hawaiian Youth.* Dissertation presented to the faculty of the School of Education Leadership Studies Department, Organization and Leadership Program in partial fulfillment of the requirements for the degree Doctor of Education, San Francisco.

Irwin, G. (2006). Voyaging and Settlement. In K. R. Howe (Ed.), *Vaka Moana—Voyages of the Ancestors*. Honolulu: University of Hawaii Press.

Iversen, R. T., Dye, T., & Paul, L. M. (1990). *Native Hawaiian Fishing Rights. Phase One: The Northwestern Hawaiian Islands, and Phase 2: The Main Hawaiian Islands and the Northwestern Hawaiian Islands*. Prepared for The Western Pacific Regional Fishery Management Council, Honolulu.

Jokiel, P. L., Rodgers, K. S., Walsh, W. J., Polhemus, D. A., & Wilhelm, T. A. (2011). Marine Resource Management in the Hawaiian Archipelago: The Traditional Hawaiian System in Relation to the Western Approach. *Journal of Marine Biology, 2011*, 1–16.

Judd, N. L. M. (1998). Lāʻau lapaʻau: Herbal Healing Among Contemporary Hawaiian Healers. *Pacific Health Dialog, 5*(2), 239–245.

Kaʻaiʻai, C. (2018, December). Personal Communication. Former coordinator of indigenous programs at the Western Pacific Regional Fishery Management Council, Honolulu.

Kahāʻulelio, A. D. (1902, February and March). Fishing Lore. In M. K. Pukui (Trans.), *Nupepa Kuokoa*. Honolulu: Bishop Museum.

Kahāʻulelio, A. D. (2006). *Ka ʻOihana Lawaiʻa—Hawaiian Fishing Traditions* (M. Puakea Nogelmeir, Ed. and M. K. Pukui, Trans.). Honolulu: Bishop Museum Press.

Kamakau, S. M. (1976). *Na Hana a ka Poʻe Kahiko* (The Works of the People of Old). Translated from the Newspaper Ke Au ʻOkoʻa by M. K. Pukui. Arranged and edited by D. B. Barrere. Bernice Bishop Museum Special Publication 61. Honolulu: Bishop Museum Press.

Kamakau, S. M. (1992). *Ruling Chiefs* (Rev. ed.). Original edition compiled in 1961. Honolulu: Kamehameha Schools.

Kawaharada, D. (Ed.). (2006). *Hawaiian Fishing Traditions*. Honolulu: Noio/Kalamakū Press.

Kirch, P. V. (1974). The Chronology of Early Hawaiian Settlement. *Archaeology and Physical Anthropology in Oceania (APAO), 9*, 110–119.

Kirch, P. V. (1984). *The Evolution of Polynesian Chiefdoms*. Cambridge: Cambridge University Press.

Kirch, P. V. (1985). *Feathered Gods and Fishhooks: An Introduction to Hawaiian Archaeology and Prehistory*. Honolulu: University of Hawaiʻi Press.

Kirch, P. V., Mertz-Kraus, R., & Sharp, W. D. (2015, January). Precise Chronology of Polynesian Temple Construction and Use for Southeastern

Maui, Hawaiian Islands Determined by ^{230}Th dating of Corals. *Journal of Archaeological Science, 53,* 166–177.

Kittinger, J. N., Teneva, L. T., Koike, H., Stamoulis, K. A., Kittinger, D. S., Oleson, K. L. L., et al. (2015). From Reef to Table: Social and Ecological Factors Affecting Coral Reef Fisheries, Artisanal Seafood Supply Chains, and Seafood Security. *PLoS ONE, 10*(8). Published 5 August 2015. https://doi.org/10.1371/journal.pone.0123856.

Kosaki, R. H. (1954). *Konohiki Fishing Rights* (Report No. 1). Honolulu: Legislative Reference Bureau.

Kupau, S. (2004). Judicial Enforcement of Official Indigenous Languages: A Comparative Analysis of the Maori and Hawaiian Struggles for Cultural Language Rights. *University of Hawaii Law Review, 26,* 495–535.

Kuykendall, R. S. (1967). *The Hawaiian Kingdom 1874–1893, the Kalakaua Dynasty.* Honolulu: University of Hawaii Press.

Langlas, C. (2003). *Kau Lāʻau and Maʻamaʻ: Traditional Hawaiian Ulua Fishing.* Pili Productions. Langlas@hawaii.edu. Hilo.

Lincoln, N. K., & Vitousek, P. M. (2017). *Indigenous Polynesian Agriculture in Hawaii.* Oxford Research Encyclopedia of Environmental Science. environmentalscience.oxfordre.com. Oxford University Press, USA. Available at http://www.ulumaupuanui.org/uploads/1/8/2/1/18219029/indigenous_polynesian_agriculture_in_hawai%CA%BBi.pdf.

Lowe, M. K. (2004). The Status of Inshore Fisheries Ecosystems in the Main Hawaiian Islands at the Dawn of the Millennium: Cultural Impacts, Fisheries Trends and Management Challenges. In A. M. Friedlander (Ed.), *Status of Hawaiʻi's Coastal Fisheries in the New Millennium* (pp. 12–107). Honolulu: Hawaiʻi Audubon Society.

MacCaughey, V. (1917). The Food Plants of the Ancient Hawaiians. *The Scientific Monthly, 4*(1), 75–80.

Maly, K., & Maly, O. (2002). *He Wahi Moʻolelo ʻohana No Kaloko me Honokōhau ma Kekaha o Na Kona—A Collection of Family Traditions Describing Customs, Practices, and Beliefs of the Family and Lands of Kaloko and Honokōhau, North Kona, Island of Hawaiʻi.* Hilo: Kumu Pono Associates.

Maly, K., & Maly, O. (2003). *Ka Hana Lawaiʻa A Me Na Koʻa O Na Kai ʻEwalu: A History of Fishing Practices and Marine Fisheries of the Hawaiian Islands* (Vols. I and II). Hilo: Kumu Pono Associates. Prepared for the Nature Conservancy and Kamehameha Schools.

McGregor, D. (2007). *Na Kuaʻaina: Living Hawaiian Culture.* Honolulu: University of Hawaii Press.

Meller, N. (1985). *Indigenous Ocean Rights in Hawaii.* Honolulu: University of Hawaiʻi Sea Grant College Program, Sea Grant Marine Policy and Law Report. University of Hawaiʻi at Mānoa, Honolulu.

Melrose, J., Perroy, R., & Cares, S. (2016). *Statewide Agricultural Land Use Baseline 2015.* Prepared for the Hawaiʻi Department of Agriculture by the University of Hawaiʻi at Hilo, Spatial Data Analysis & Visualization Research Lab, Hilo.

Newman, S. (1970). *Hawaiian Fishing and Farming on the Island of Hawaiʻi: A.D. 1778.* Honolulu: State of Hawaiʻi, Department of Land and Natural Resources.

Niʻihau Cultural Heritage Foundation. (2018). *The Language of Niihau.* Available at http://www.niihauheritage.org/niihau_language.htm.

NOAA Fisheries Pacific Islands Fisheries Science Center. (2018). *Hawaii Bottomfish Heritage Interviews. Infographic* (various technical products forthcoming). Honolulu. Programmatic blog entry available at https://www.fisheries.noaa.gov/feature-story/hawaii-bottomfish-heritage-project.

Nogelmeier, P. (2006). *Ka ʻOihana Lawaiʻa—Hawaiian Fishing Traditions* (M. Puakea Nogelmeir, Ed. and M. K. Pukui, Trans.). Honolulu: Bishop Museum Press.

Ogawa, M. (2015). *Sea of Opportunity: The Japanese Pioneers of the Fishing Industry in Hawaiʻi.* Honolulu: University of Hawaii Press.

Oliver, D. (2002). *Polynesia in Early Historic Times.* Honolulu: Bess Press.

Papa Ola Lokahi. (2008). *Chronology of Events Relating to Traditional Hawaiian Healing Practices Since 1985.* Honolulu. Available at http://www.papaolalokahi.org/images/CHRONOLOGY-of-EVENTS-RELATED-TO-TRADITIONAL-HEALING-2015-Dec.pdf.

Paulo, W. (2002, January). Personal Communication. Miloliʻi, Hawaiʻi.

Poepoe, K. K., Bartram, P. K., & Friedlander, A. M. (2003). The Use of Traditional Hawaiian Knowledge in the Contemporary Management of Marine Resources. In *Putting Fishers' Knowledge to Work* (pp. 328–339). Vancouver, Canada: Fisheries Centre Research Report, University of British Columbia.

Schug, D. (2001). Hawaii's Commercial Fishing Industry: 1820–1945. *The Hawaiian Journal of History, 35,* 15–34.

Scobie, R. (1949). *The Technology and Economics of Fishing in Relationship to Hawaiian Culture.* Thesis completed in partial satisfaction of the requirements for the Master's degree in economics, London School of Economics, London.

Severance, C., Franco, R., Hamnett, M., Anderson, C., & Aitaoto, F. (2013). Effort Triggers, Fish Flow, and Customary Exchange in American Samoa and the Northern Marianas: Critical Human Dimensions of Western Pacific Fisheries. *Pacific Science, 67*(3), 383–393. In Special Issue of *Pacific Science*—Human Dimensions of Small-Scale and Traditional Fisheries in the Asia-Pacific Region (J. Kittinger & E. W. Glazier, Eds.).

Sibert, J. (2018). *Assessing a portion of the Pacific thunnus albacares stock: Ahi in the main Hawaiian Islands*. ArXiv pre-print archive. Available at https://arxiv.org/pdf/1702.01217.pdf.

Silva, N. K. (2000). He kanāwai e ho'opau i na hula kuolo Hawai'i: The Political Economy of Banning the Hula. *Hawaiian Journal of History, 34,* 29–48.

Silva, N. K. (2004a). *Aloha Betrayed—Native Hawaiian Resistance to American Colonialism*. Durham and London: Duke University Press.

Silva, N. K. (2004b, June). The Importance of Hawaiian Language Sources for Understanding the Hawaiian Past. *ESC. Reader's Forum, 30*(2), 4–12.

State of Hawaii, Department of Business, Economic Development, and Tourism. (2000, May). *Construction and Hawaii's Economy*. Economic Analysis Division. Honolulu.

Stecyk, C. (1993). *Ho'olina*. Patagonia Ad. *Surfer's Journal, 2*(4), 128.

Stout, M. A. (2012). *Native American Boarding Schools*. Landmarks of the American Mosaic (series). Greenwood Publishers. An Imprint of ABC-CLIO, LLC. Santa Barbara, CA.

Tabrah, R. (1984). *Hawaii: A History*. New York: W. W. Norton.

Teneva, L. T., Schemmel, E., & Kittinger, J. N. (2018). State of the Plate: Assessing Present and Future Contribution of Fisheries and Aquaculture to Hawai'i's Food Security. *Marine Policy*. Accepted April 18. https://doi.org/10.1016/j.marpol.2018.04.025.

Titcomb, M., & Pukui, M. K. (1951). Native Use of Fish in Hawai'i. Memoir 29. *Supplement to the Journal of the Polynesian Society* (Installment No. 1), 1–96.

Townsend, C. K. M. (2014). *Impacts of Hawaiian Language Loss and Promotion Via the Linguistic Landscape*. A dissertation submitted to the Graduate Division of the University of Hawaii at Manoa in partial fulfillment of the requirements for the degree of Doctor of Public Health, Honolulu.

Valeri, V. (1985). *Kingship and Sacrifice—Ritual and Society in Ancient Hawaii* (P. Wissing, Trans.). Chicago: University of Chicago Press.

Vaughan, M. B., & Ayers, A. L. (2016). Customary Access: Sustaining Local Control of Fishing and Food on Kaua'i's North Shore. *Food, Culture & Society, 19*(3), 517–538.

Vaughan, M. B., & Vitousek, P. M. (2013). Mahele: Sustaining Communities Through Small-Scale Inshore Fishery Catch and Sharing Networks. *Pacific Science, 67*(3), 329–344. In Special Issue of *Pacific Science*—Human Dimensions of Small-Scale and Traditional Fisheries in the Asia-Pacific Region (J. Kittinger & E. W. Glazier, Eds.).

Yuen, H. S. (1979). A Night Handline Fishery for Tunas in Hawaii. *Marine Fisheries Review, 41*(8), 7–14.

Yuen, H. S. (2016, August). Personal Communication. Oral history interview conducted as part of the Voices of the Fishery Science Centers project undertaken by the U.S. Department of Commerce, NOAA Fisheries, Northeast Fisheries Science Center, Honolulu.

4

Applying Tradition to the Contemporary Resource Management Process

4.1 The Resurgence of Native Hawaiian Society and Culture

During the mid-1970s, after decades of quiet resistance to oppressive aspects of foreign society and culture, indigenous activists openly confronted the colonial powers that claimed ownership of the Hawaiian Islands. The movement began in earnest with a series of assertive actions to reclaim the island of Kahoʻolawe from U.S. military forces, which had used and damaged the land and island ecosystems through munitions testing and development since World War II. The concurrent revival of ancient traditions and widespread renewal of Polynesian identity constituted a cultural renaissance that continues today (Inafuku 2015: 35–36):

> From 1976 through the 1990s, a general cultural resurgence spread through [Hawaiʻi] as the number of hālau hula (schools for instruction in traditional Hawaiian dance and chant) (Pukui and Elbert 1986: 88) increased, Hawaiian language and music gained popularity, and traditional Hawaiian healers began training a new generation in lāʻau lapaʻau (medicine) (Pukui and Elbert 1986: 189). Of international significance,

© The Author(s) 2019
E. W. Glazier, *Tradition-Based Natural Resource Management*,
Palgrave Studies in Natural Resource Management,
https://doi.org/10.1007/978-3-030-14842-3_4

Native Hawaiians revived traditional navigational skills and showed the world the ability to travel the Pacific Ocean without the aid of instruments. After decades of purposeful cultural suppression by American government leaders, Native Hawaiians regained passion and pride for their culture (McGregor 2007: 276–277). Inspired by this renewed cultural pride and appalled by the indifferent desecration of Kahoʻolawe's natural resources and cultural sites, George Helm and a group of young Native Hawaiians founded the Protect Kahoʻolawe ʻOhana ("PKO") in 1976 (Osorio 2014: 143). Dedicated to stopping [test and practice] bombing on Kahoʻolawe and reclaiming the island for the Native Hawaiian people, Helm and PKO became an integral part of the growing political and cultural resurgence among Native Hawaiians. Helm used his gift for music to lead a grassroots campaign in raising awareness of the United State's destruction of Kahoʻolawe and to inspire organized efforts to protest that destruction. PKO embraced the environmental, political, and spiritual meaning of aloha ʻāina (love of the land). Taken deeper, aloha ʻāina represents a duty to organize and rally the people around Native Hawaiian rights and sovereignty in order to achieve the political standing necessary to protect the land. Helm explained his commitment to aloha ʻāina regarding Kahoʻolawe in the following way: "I am a Hawaiian and I've inherited the soul of my kūpuna (ancestors). It is my moral responsibility to attempt an ending to this desecration of our sacred ʻāina, Kohe Mālamalama O Kanaloa (Kahoʻolawe), for each bomb dropped adds further injury to an already wounded soul."

Thus started the ongoing movement toward self-determination and publicly renewed emphasis on indigenous culture in the Hawaiian Islands. Efforts to protect Kahoʻolawe, slowed by various political and bureaucratic hurdles, were eventually successful. Bombing was halted through litigation in 1990; the island was deeded from the U.S. Navy to the State of Hawaiʻi in 1993; and the Kahoʻolawe Island Reserve Commission assumed administrative control in 2003. The Commission retains overarching responsibility for stewardship of the island and its natural resources until it can be transferred to a Native Hawaiian entity for long-term management.

The 1970s were a time of outright resistance among certain indigenous groups in North America, Polynesia, Australia, and elsewhere around the world. What were once considered traditional societies fated to disintegrate in a world of capitalist ideals, pressures, and processes have, in many regions, proven highly resilient. In his insightful treatise on macro-social forces of change and continuity of culture among indigenous North Americans, Champagne (2007: 9) presents two competing perspectives on this matter. The first is a fatalistic view the author envisions as being erroneously embedded in contemporary scientific thought and theory:

> Except for some anthropologists and environmentalists, indigenous communities are often considered doomed to destruction at worst or submerged within nation states at best. Current theories of globalization and world systems suggest that indigenous communities will be drawn into dependent, exploited relations...and remain backwaters of economic and social organization. Melting pot theories and multicultural views leave little room for indigenous rights, claims to territory, and political and cultural autonomy. The image of the vanishing [Native] continues to find subtle sway in intellectual and popular circles.

The second perspective is quite different, more positive in nature, and more empirical than prophetic. It also makes clear that the individual and collective agency needed to effect the kinds of political changes that can truly benefit indigenous societies continues to be applied around the globe. In short, a distinctive individual and group identity as indigenous tends to preclude wholesale capitulation to the values of impinging societies and cultures. Of this Champagne (2007: 9) writes:

> During the 1980s and 1990s, many indigenous groups around the world increasingly found their voice in national and international forums. Land rights and political recognition in New Zealand, Australia, and Canada have taken major positive turns in favor of indigenous groups when compared to conditions that existed several decades ago. Native communities weathered by the colonial experience seek renewal, demand rights to self-government and cultural autonomy, and increasingly negotiate relations with nation-states. Many Native communities and identities have not disappeared or assimilated, contrary to the predictions of evolutionary, integrative, and multicultural theories. Native communities promise to be enduring participants in international and national affairs in coming centuries. (Dippie 1982; Ramirez 1998)

Clearly, the second perspective captures the situation of the Kanaka Maoli. As discussed in previous chapters, Native Hawaiians have, over the past 240 years, proven that disappearance of the society and culture is in no way inevitable or even likely. Generations of Hawaiians have continued to practice and evolve core elements of the culture despite the ongoing arrival of foreign populations of continually expanding size, political influence, and socioeconomic impact. Even the intermarriage of Native Hawaiians with locally situated persons of other ancestries has not diminished the culture. To the contrary, in conjunction with persistent attention to customs and traditions, intermarriage has reified expression of Native Hawaiian culture by expanding the size of the population of persons with indigenous roots in the islands.

While the speaking of Hawaiian and various cultural practices were suppressed by haole (foreign) missionaries in the nineteenth century, and by haole educators into the twentieth, such actions ultimately diminished—partly through resistance on the part of Kanaka Maoli, and partly through the effects of demographic and cultural trends that lend to increasing political power among the indigenous population. For example, persons solely of European descent now constitute a minority island population when measured against a highly admixed overall population, and the ethnocentric concepts that Caucasians once openly sought to advance are widely seen as part of an ignorant past. Unfortunately, the recent resurfacing of white supremacist culture (Daniels 2018) and ongoing struggles between indigenous and non-indigenous populations on the North American continent (Loppie et al. 2014; Blendon et al. 2017) clearly indicate that appreciation of indigenous society is less than universal as we approach the third decade of twentieth century America.

In any case, haole perspectives are often largely irrelevant in the context of the dominant local culture that is characteristic of many Hawai'i communities today. Kama'āina (long-time residents) are steeped in local knowledge, customs, and linguistic expertise, while visitors and short-term residents generally possess limited understanding of island life and the rich history of its original colonists. Moreover, Native Hawaiian ancestry, the ability to speak the language, and capacity and readiness to practice the culture are sources of status, pride, and self-identity among island residents of Native Hawaiian ancestry. This is particularly significant since the number of Native Hawaiians and part-Native Hawaiians living in the state is growing rapidly, more than tripling in size between 1970 and the present. The figure now approaches 300,000 persons or more than 21% of the state's total population in 2013 (U.S. Census Bureau 2015), with projections holding that more than 500,000 persons of Native Hawaiian ancestry will reside in the state by 2045, and 675,000 by 2060 (Kamehameha Schools 2014; Goo 2015).

In sum, a now rapidly expanding population and collective resistance to impinging social forces mean that Native Hawaiians will exert increasingly profound sociopolitical influence across the islands in the years to come. Of note, in December 2018, the U.S. Commission on Civil Rights reversed previous opinions and now supports self-determination and indigenous-based governance among Native Hawaiians (U.S. Commission on Civil Rights 2018: ES-10). This is not necessarily suggestive of a contented future, however, since people and corporations from outside Hawai'i will continue to bring conflicting cultural values and the intent to further develop available land for individual or corporate use and economic gain. Inasmuch as this process is contrary to the ongoing struggle for land, rights, and collective well-being among the descendants of the original inhabitants and their long path to the cultural and demographic resurgence of the present day, problems seem unavoidable (Glazier 2007: 131–134).

The path to the present state of resurgence of Native Hawaiian society and culture has been long and challenging. The cultural renaissance and quest to amend basic injustices of the past have involved many Kanaka Maoli, some still living and others who have now passed on. Such efforts include but are by no means limited to: the aforementioned establishment of Hawaiian language immersion schools across the state, many scores of legal cases seeking to preclude land development and associated impacts on Native Hawaiian society and the natural environment, and major policy debates that could improve the rights and opportunities of indigenous Hawaiians across the islands. The ongoing movement toward self-determination continues to play out in a variety of ways across various social and political arenas.

The burden of effecting social and political change generally falls on Native Hawaiians themselves—despite the fact that the illegality of the annexation and overthrow of the Hawaiian monarchy are seen by many activists, academics, and others as having caused a cascade of profound events and actions that had and have no real basis in international law (Sai 2011: 121). These include: the granting of statehood; the ongoing acquisition and holding of lands and resources by haole entrepreneurs, corporations, and government entities (Novak 1993: 191–203); the continuing use and occupation of lands for military purposes; the displacement of Kanaka Maoili from their ancestral ahupua'a through ongoing economic processes; and the continuing denial of inherent Native Hawaiian sovereignty by external agencies and entities.

The perspectives of Kanaka Maoli on such matters, and the steps needed to address these and associated problems of contemporary life in the islands vary extensively—as would be expected of any complex society in the modern age. Indigenous Hawaiians speak for themselves on such matters. Nevertheless, there is widespread agreement that much injustice has transpired and will be righted. This is not merely a complaint—it is rather a collective proclamation that many thousands of Kanaka Maoli and others remain dedicated to addressing in the years to come. This profound intention tends to impassion 'aha (assemblies, meetings) held by contemporary Native Hawaiians, especially when the discussion at hand has the potential to affect hanauna (generations) to come.

4.2   Traditional Use of Natural Resources in Modern Hawai'i

The use and management of natural resources, and perhaps especially living marine resources, are vital contemporary topics with implications for the future of indigenous culture and society in the Hawaiian Islands. It is not an exaggeration to state that opportunities to fish, hunt, gather, and grow food for purposes of direct consumption, giving, sharing, customary trade, and other social and economic interactions have profound and at times even life and death implications for certain 'ohana (families). This relates both to the dietary-nutritional production value of such activities, and to the profound benefits that can stem from participation in traditional ways of life, including self-identity, social meaning, and even the will to live.

With regard to food production value, wild and self-cultivated foods provide sources of nutrition which, for many, complement goods and products purchased with monies generated through the various economic inputs typical of twenty-first century Hawai'i. Such food can be vitally important in certain settings. This is noted, for example, by Teneva et al. (2018: 33):

> Non-commercial fisheries ... play a major role in the subsistence economy, providing millions of meals per year directly to households (Grafeld et al. 2017; Kittinger et al. 2015; Glazier et al. 2013; Severance et al. 2013). A substantial percentage of this seafood is shared via social kinship networks (Kittinger et al. 2014, Vaughan et al. 2013). This food distribution system plays a major role in the informal economy, is highly prevalent, and supports food security in communities across the [Hawaiian] archipelago. Ensuring the persistence of these activities should be a key policy priority in order to support the cultural traditions and food security of local communities.

Of the largest resident ancestral population groups identified in the 2015 American Community Survey (ACS) 5-Year Estimates (Caucasians, Filipinos, Japanese, Native Hawaiians, and Chinese), Native Hawaiians report the highest rates of poverty in Hawai'i for both individuals and families, with 6610 families (12.6% of the population) and 45,420 individuals (15.5% of the population) living below the poverty level. By comparison, statewide rates of poverty are 7.7% for all resident families, and 11.2% for all resident individuals. Given relatively higher rates of poverty and traditional use of wild food resources, the importance of such foods is proportionately greater for Native Hawaiians especially important in island areas where economic challenges are particularly acute and/or where the indigenous population is extensive. For instance, Matsuoka et al. (1994) found that wild foods comprised 38% of the indigenous diet on the predominately Native Hawaiian island of Moloka'i, a figure that is very likely approximated today. The town of Wai'anae on the island of O'ahu, a predominately Native Hawaiian community with an estimated year 2015 family poverty rate of 26% provides a more urban example as the pursuit, distribution, and consumption of seafood remain vitally important aspects of life here (Glazier et al. 2013).

Lawa'ia (fishing), mahi'ai'ana (farming), hahai holoholona (hunting), and hō'il'ili (gathering) are pivotally important traditional practices among Native Hawaiians. These provide vital connections among the person, the surrounding island and ocean environment, and one's cultural roots. Harvested foods are often used for ceremonial or celebratory purposes, furthering social interaction and cohesion within and across 'ohana holo'oko'a (extended family) and kaiāulu (community). Because hunting, fishing, farming, and related activities typically involve multiple participants who fulfill culturally designated roles and who possess ecological knowledge that is developed and refined across generations, the activities also facilitate laulima (literally many hands, or the cooperation needed to achieve a collective goal). This is readily observable in many island settings, at least for those with an interest in such matters. Glazier (2007), for instance, observed numerous instances of cooperative social interaction in the context of nearshore and off-shore fishing activities in remote Native Hawaiian communities along the Kona (leeward) side of Hawai'i Island during the 1990s and early 2000s. Some such instances revealed the persistence of a watershed-oriented system of land use and management, and the important roles that fishermen and farmers continue to play in rural Hawaiian society:

> Michael Ho'omakoa (a pseudonym but a common mix of English and Hawaiian names that is illustrative of coexisting cultural systems in the islands) agrees to take me fishing in the wa'a pa or three-board canoe. We prepare to leave from the beach along a cove fronting a little village on the Big Island. Five men and four boys assemble at the beach and confer quietly, elbows resting on the back of Terry's pickup truck. Michael lays out this sunny day's plan. He speaks with authority like a chief, and all are silent until he finishes his sentences. At his command, all nine help get the canoe past the surf zone. But only four remain in the canoe to go fishing. The rest drive or walk off to work or play, having stated they will return later in the day to assist in bringing the boat back to the beach.

Terry is going back up to his mauka [mountain] homestead to tend to pigs and cattle. But Michael assures him he will get a share of the catch later in the day. There is exchange of goods and close social relations between residents of sea and mountain portions of this valley [ahupua'a], as in old Hawai'i. (Glazier 2007: 128)

Indigenous traditions in Hawai'i abound with social meaning for those involved. The traditional baby lū'au (feast), for instance, is convened by native and local families throughout the islands to celebrate the child's one-year anniversary—an especially happy occasion given the challenges of attaining this age here in the not so distant past. Kamehameha Day, Prince Jonah Kuhio Kalanianaole Day, and other Hawaii-specific holidays bear particular cultural significance and typically occasion pā'ina (celebratory feasting), as do weddings, funerals, graduations, and other life events. The various holidays celebrated on the continent also induce social gatherings in the islands—Christmas, Easter, July 4th, and so on. Large celebrations and ceremonies very often occur in seaside parks, with numerous large tents protecting attendees from the sun and occasional tradewind shower. Shoreline fishing and the regal Hawaiian sports of surfing and canoe paddling often occur nearby, as they have for centuries. Deep respect for the Pacific Ocean and its great power is a strong cultural norm among Native Hawaiians.

The pā'ina is central to Hawaiian culture, involving: extensive kōkua (assistance, cooperation) among all in the extended family; much talk story (conversation); a good dose of ho'omāke'aka (humor); Hawaiian music, of which there are numerous rich and extensive genres; various libations; and the preparation and consumption of traditional foods. These typically include: dishes such as poi—a starchy past made from the corm of kalo; poke—raw cubed fish served in a marinade of shoyu, shallots, limu (seaweed) and/or other ingredients; laulau—various meats or certain seafoods such as he'e (octopus) wrapped in leaves and cooked in the imu (underground oven); lomi-lomi salmon (a chilled dish prepared by massaging the fish, typically imported from Alaska or the Pacific Northwest, with ingredients such as tomatoes, Maui onions, salt, and red pepper); and kālua pig, specially prepared in the imu after capture in the rugged island mountains. Such foods and their careful preparation are essential among those with an eye to custom and tradition.

Attendees bring specific dishes to a given event and inevitably leave with a general array of leftovers.

An informal culture of sharing resources is a common aspect of social life in rural parts of Hawai'i and in local-urban neighborhoods alike—such as in Papakōlea, a Native Hawaiian neighborhood located immediately mauka (toward the mountains) above densely populated Honolulu. Wild foods harvested from mountains and sea in and above even this urban region are important components of an economy in which members of various 'ohana interact regularly to share: childcare duties, the costs of various pā'ina, rides across town, cash, tutelage of keiki (kids), fishing trip costs, and so on. Such social and economic ties are, of course, forged in communities around the world, but always with unique cultural, place-specific, and situational meaning. Among Native Hawaiians, the sharing of resources across the extended family and kaiāulu (Vaughan 2018) is emphasized in local settings where macro-economic changes are fairly recent and where the terms of an older subsistence economy still linger. Even the duties and costs of taking in an orphan or other person who needs a home are assumed by 'ohana around the islands through an age-old practice called hānai (to raise, rear, nourish, sustain—in this case, a person without a home).

Macro-social Challenges

Informal reciprocity and strong social relationships between families characterize social and economic life in many island communities today. But in a setting where the past is both culturally valued and painful, and where core indigenous values of cooperation and sharing are often confounded by modern societal emphasis on individual achievement, a state of individual and collective well-being can be elusive. Notably, a relatively high percentage of Native Hawaiians experience significant health problems (Galinsky et al. 2014), high rates of substance abuse (Mokuau 2002: 583–584), short life expectancies (Duponte et al. 2010: 2), persistent poverty (Hostetler 2014), inordinately high rates of incarceration and disparate treatment in the criminal justice system (State of Hawaii 2010), and a rate of homelessness that is unsurpassed in the islands. Homelessness is discussed by McDougall (2010: 52), who notes how Native Hawaiians persevere even as they are forced to move from island beaches and parks where alternative forms of shelter and other life necessities can be found:

> Homelessness is a key issue for Kānaka Maoli, as the majority of the homeless population [in Hawai'i] is of Kanaka Maoli descent, a fact that highlights the colonial dispossession of our 'āina. Without houses, many homeless families must live on beaches, where they have access to a source of food, showers, and spirituality through our familial connection to the ocean and the land. Invariably, evictions from beaches become important sites demonstrating this colonial dispossession. However, they also exemplify Kanaka Maoli resistance to further displacement and our defiant survival by returning to the land for our sustenance.

The collective shock induced by persistent large-scale disruption to a given society, such as occurred in the Hawaiian Islands between the eighteenth and twentieth centuries, is increasingly thought to have generated inter-generational impacts of particularly long duration. Given trying and persistent histories of oppression, the concept of "historical trauma" is now being applied by social scientists, psychologists, and social workers to help explain and address inordinate social and public health challenges among Alaska Natives, Native Americans, Pacific Islanders, and Native Hawaiians. Sotero (2006: 94) and supporting literature provide insight into the phenomenon, with the author asserting that "populations historically subjected to long-term mass trauma exhibit a higher prevalence of disease even several generations after the original trauma occurred (Danieli 1998; Brave Heart and DeBruyn 1998; Duran and Duran 1995; Leary 2005)." This is reiterated with specific regard to the Hawai'i case by Cook et al. (2003: 10) who assert that "key [historic] incidents of dramatic and sudden social change reveal traumatological mechanisms sufficient in size that they may account for health disparities faced by [Native] Hawaiians today." Similarly, Mokuau et al. (2016: 2) state that "although the overthrow of the Kingdom of Hawai'i occurred more than a century ago, historic loss of population, land, culture, and self-identity have shaped the economic and psychosocial landscapes of Hawai'i's people and limits their ability to actualize optimal health."

A growing literature also indicates viable solutions to problems stemming from historic oppression of indigenous people in what is now the United States. For example, Brown-Rice (2013: 117–130) provides evidence-based discussion of cross-generational transmission of trauma resulting from historical losses among Native Americans on the continent, with insights into etiology and intervention that are highly relevant to the Hawai'i case. The author emphasizes the need for health practitioners to understand both the varied post-contact histories of indigenous societies, and group-specific traditional strategies for addressing contemporary social and public health challenges (ibid.: 127). Notably, the post-colonial experiences and perspectives of Native Hawaiians vary, as do preferred solutions to various challenges. For this reason, group deliberation on social issues of importance to Native Hawaiians ideally involves broad representation of indigenous experiences, needs, and interests.

Reconnecting to Land and Sea

Reconnection to traditional stewardship and use of natural resources is increasingly thought to aid in diminishing social and public health problems presently being experienced by Native Hawaiians (State of Hawai'i 2010: 24; Perez 2016; Liu 2005: 90–92) and by American Indian and First Nation tribes on the continent (Brave Heart 2003; DeBruyn et al. 2001; Lechner et al. 2016). Mokuau et al. (2016: 6) suggest that culturally sensitive health care strategies and reengagement with traditional culture and traditional-holistic perspectives on health and well-being are essential for addressing contemporary problems stemming from historical trauma, including those associated with loss of the "original agricultural and aquacultural way of life due to urbanization (Liu and Alameda 2011), the replacement of the Hawaiian language with English in legal and educational settings (Liu and Alameda 2011), limited access to Native foods due to cost and restrictive land use policies, and relegation of native culture to Western depictions of Polynesian culture for tourist advertisements (Kana'iaupuni and Malone 2006)." With regard to prospective means for addressing health problems and achieving health equity in a modern indigenous context that is so clearly linked to the past, Mokuau et al. (2016) asserts the utility of balance and holism as valued by cultural practitioners themselves:

> Native Hawaiians describe cultural knowledge and practice to be of main importance (McMullin 2005). They believe that a balanced system that integrates all aspects of the self (biological, psychological, social, cognitive, spiritual) with the world (individual, family, community, environment) brings about optimal health (McMullin 2005; Mokuau 2011). Native Hawaiian cultural values and beliefs are organized around the collective relationships of the family, community, land, and the spiritual realm. (pp. 4–5)

Mokuau (2002: 585–586) also describes human service interventions appropriate for persons of Native Hawaiian ancestry. One such approach is termed "aloha 'āina," or caring for the land. According to the author, such interventions should reflect the Native Hawaiian cosmography that people originate from island and ocean and are stewards and beneficiaries of the resources of land and sea. She states that "the relationship and reciprocity of land and its caregivers are best experienced through cultural immersion activites and programs that require participants to live with the land" (p. 586) and that such programs can alleviate problems such as substance abuse. For example, in describing a non-profit *aloha 'āina* program that facilitates hands-on cooperative work in kalo fields on the Big Island, Mokuau (2011: 11) states that:

> Since *Lo'i Kalo* derives from our ancestral past, there is a permeation of cultural values and practices. Working in the *lo'i kalo* exemplifies having a "sense of place." Oneha (2001) suggests that a sense of place relates to being rooted to an ancestral place, and experiencing *pono* (balance), *mana* (spiritual power), and *kuleana* (responsibility). The responsibility of nurturing our relationship with the land also coincides with our relationship with others. Working with others in the *lo'i* generates values of the collective, such as *kokua* (helping), *laulima* (cooperation) and *lokahi* (unity) [all of which can diminish a sense of isolation among those seeking relief from various social problems].

The success of such programs requires unencumbered opportunities for Kanaka Maoli to consistently participate in the traditional ways of life that formerly contributed, and for some residents continue to contribute, to well-integrated 'ohana and island communities. Such lifeways definitely include customary knowledge, pursuit, harvest, sharing, and consumption of living marine and terrestrial resources and the products of small-scale agriculture. Indeed, coupled with varying ways and degrees of participation in the modern economy, traditional use of the natural environment remains a natural and profound dimension of indigenous life in Hawai'i and other parts of Oceania—a dimension that indigenous people across the region will undoubtedly continue to advocate in natural resource policy decisions for years to come.

Constraints on Traditional Use and Management of the Marine Environment

Indigenous participation in the management of marine fisheries in the present-day has been complicated by the fact that natural resources and the rights of Native Hawaiians to pursue, use, and manage such resources in traditional fashion continue to be threatened by expansion of non-indigenous populations and the environmental problems and sociopolitical constraints that accompany such growth in small island settings of the Pacific. Threats to cultural and natural resources are now chronic in the Hawaiian Islands. Arable land is at a premium due to widespread development pressure (Melrose et al. 2016: 6); Native Hawaiians must litigate to protect their 'iwi kūpuna (sacred ancestral bones) (Baldauf 2014); Hawai'i is home to more endangered species than any other state (Underwood et al. 2013: 1); nearshore fisheries production has diminished substantially over the past century (Friedlander et al. 2015); and coral reef fish assemblages and ecosystems are increasingly impacted by human activities (Williams et al. 2015; Rodgers et al. 2015; Wedding et al. 2018; Silbiger et al. 2018).

With specific regard to the availability of marine resources for consumption, sale, sport, and reciprocal and customary exchange, challenges are not new in Hawai'i. In fact, problems with fish stocks were observed here as early as the late nineteenth century. For example, Titcomb and Pukui (1951: 11) note that in Hawaiian newspapers of the early 1900s, it was often asked, "Why are the fish so scarce and prices so high?"

Populations of pelagic, nearshore, and reef-associated fish species are subject to many physical environmental factors and processes that limit and/or enhance fisheries productivity. But it is clear that unfettered human interaction with living marine resources in general can ultimately limit the abundance and/or alter the distribution of certain species through overharvesting, and impacts to marine and terrestrial ecosystems that support such resources. The recent work of McCoy et al. (2018) elucidates both the consumptive-dietary (and recreational) importance of reef-asociated fish species to modern-day island residents, and the human pressures that continue to affect such relatively static populations of fish. Meanwhile, the movement and migratory patterns of tuna and other important open ocean species are only now becoming understood (e.g., Rooker et al. 2016: 277), and thus the effects of human activities on such species are not as clearly known as for many relatively less mobile species that can be studied in situ. Yet, contemporary fishery scientists such as Itano and Holland (2000: 54–55, 60) and Wells et al. (2012: 194) provide evidence suggesting the presence of tuna sub-stocks with affinity to Hawai'i. Although tuna are not presently considered overfished in the region, logically, such stocks are subject to the effects of fishing and other human activities emanating from the Hawaiian chain. In any event, the dynamic interplay between physical-environmental factors and processes and those stemming from human actions and processes is a critically important subject of fishery scientists worldwide, many of whom increasingly recognize the need to understand marine ecosystems in ways that more fully account for the roles and effects of human society, including the effects of reimplementing traditional management strategies (Glazier 2011).

As discussed in previous chapters, harvest practices in Old Hawai'i were in great measure carefully regulated, and typically in a way suited to specific ecological conditions and opportunities in and adjacent to a given ahupua'a or moku (traditional district). Such place-specific regulation undoubtedly played a key role in the successes achieved by Hawaiians prior to contact with Europeans. But subsequent to the arrival of haole explorers and settlers, and the many challenges that ensued, marine resources as a whole began to be treated less as to-be-conserved sources of food for indigenous families and more as commodities for transaction in the marketplace, thereby increasing pressure on a base of finite resources. This point is made by Jokiel et al. (2011: 11) and supporting authors who offer social explanations for diminishing nearshore and reef-associated fish stocks as the decline was first noted during the latter part of the nineteenth century:

"...between 1898 and 1905, detailed reports on the condition of the fisheries and management recommendations based on commercial values of catch were prepared by the U.S. Fish Commission. These data provide an important baseline that has been used to document an 80% reduction in coastal fish catch between 1900 (1,655,000 kg) and 1986 (285,000 kg). ... Following the overthrow of the Hawaiian kingdom and annexation to the United States in 1898, fisheries management was delegated to various government agencies. As was the case with colonial powers throughout much of Oceania, traditional fishing rights were systematically extinguished in the name of the discredited "freedom of the seas" concept and because such customs prevented newcomers from expropriating the islanders' resources (Johannes 1982). Ocean tenure practices based on regulation of fisheries through control of fishing rights were replaced by unlimited entry [open access], often ... [leading to] eventual resource depletion through overharvesting. The traditional system based on cooperation for the good of the community was slowly replaced by commercial forces and competition to benefit the individual. The subsistence-based, locally governed economy was converted to a cash-based economy controlled by remote global market demand. As time progressed, technology provided refrigeration and more efficient fishing gear, further accelerating the shift from subsistence to profit-based economies. A dramatic decline in Hawaiian fisheries stocks and fishery production occurred during the period of commercialization of fisheries. (Maly and Maly 2003)

Unfortunately, the declines noted of the late nineteenth century continued into the late twentieth—as discussed, for instance, in the *Findings of the Peoples' International Tribunal on the Rights of Indigenous Hawaiians* (Churchill and Venue 2004: 710). Compiled some 25 years ago, the tribunal relates social-environmental changes to the diminishing productivity of Hawai'i's fish ponds and nearshore fisheries in an expansive text that describes a large suite of problems confronting Native Hawaiians a century after the overthrow of the Hawaiian monarchy (Fig. 4.1):

Fig. 4.1 At Heʻeia on the island of Oʻahu, a Native Hawaiian community group works to restore an ancient fishpond to produce fish for consumption and to educate children and the adjacent community (Photo courtesy of the Western Pacific Regional Fishery Management Council)

> While it is true that traditional fish ponds have not as yet completely dis-appeared—the tribunal visited a functioning one of approximately 50 acres—it is equally true that the great bulk of them have been destroyed as non-natives bought up prime beachfront properties throughout the islands. For example, in 1901, Oʻahu had over 100 fish ponds. By 1993, there were only two remaining. ... Further, the future of those few which remain is anything but secure. Not only are Kanaka Maoli property rights with respect to their ponds left unsecured, but concentrations of inshore pollutants ranging from raw sewage to waste oil and PCBs threaten their viability in supporting populations of fish and other marine life.

Based on a reading of Friedlander and Brown (2004), and more specif-ically of Lowe (2004: 42), the condition of coastal and nearshore eco-systems continued to worsen at the start of the twenty-first century. The author describes numerous factors and processes that conspire to impact the productivity of fisheries conducted near most of the main islands.

All such constraints ultimately relate in some way to increases in the size of the human population residing in or visiting the Hawaiian Islands and the various ecological burdens people now generate in island settings of the Pacific:

> Cumulative and collateral impacts of cultural change and habitat degradation on inshore fisheries are poorly understood but have clearly occurred within the past century (Pooley 1987; Shomura 1987; Harman and Katekaru 1988; Maragos and Grober-Dunsmore 1999). In addition to overfishing, some of the mechanisms impacting inshore stocks and their habitat include reduction of stream productivity, the food source for many inshore species and their prey due to reduction or loss of flow (Nakasone 1995; Chong 1996); introduction of competing and predatory alien species (Uchida and Uchiyama 1986; Hunter and Evans 1995; Friedlander et al. 1997, 2002; USGS 1999; Englund et al. 2000; Yamamoto and Tagawa 2000); disruption and congestion of spawning and nursery areas (Kaneohe Bay Master Plan Task Force 1992; Lowe 1996; Commission on Water Resource Management 1995); nutrient loading, eutrophication, and parasitism favoring growth of opportunistic native and alien species (Smith et al. 1981; Font et al. 1996; Chun Smith 1994; Englund and Filbert 1996); burying and overgrowth of the reef environment (Ferguson, Wood and Johannes 1975; Evans et al. 1986; Hunter and Evans 1995); and changes in other habitat related parameters within watersheds. The effects of all these impacts on the ecological balance (herbivores versus carnivores, diversity and relative species abundance, filter feeders versus other elements of marine communities, etc.) can cause diverse collateral effects, among other things prolonging resource recovery from natural storm events. (Grigg 1972, 1994, 1995; Ferguson Wood and Johannes 1975; Kaufman 1986; Laws and Allen 1996)

Despite ongoing efforts to improve the status of Hawai'i's marine and terrestrial environs (e.g., see Keala et al. 2007), the prospective benefits of interaction between Kanaka Maoli and the surrounding marine and island environments are threatened by contemporary social-environmental problems (see NOAA Coral Reef Conservation Program 2018). This has serious implications in an age when an expanding indigenous population and new generations of persons seeking to revive traditional culture demand access to healthy island and marine ecosystems and resources. Without access to healthy marine and terrestrial ecosystems, Native Hawaiians can neither effectively practice core aspects of their culture nor acquire traditional foods for consumption, sharing, trade, and other uses. But with some 1.4 millions persons residing in the Hawaiian Islands in 2017 (Fogelman 2018; U.S. Census Bureau 2018), and a record 8.9 million visitors in 2015 (Chun et al. 2016: 2), acres for planting and points of shoreline access are increasingly scarce, pollution and alteration of physical habitats are worsening, and resource use conflicts are commonplace. Such problems are especially acute on the island of O'ahu where the number of residents now approaches 990,000 persons, with a population density of some 1600 persons per square mile in 2010 (U.S. Census Bureau 2017). Moreover, large-scale problems such as sea-level rise, shifting climate regimes, and ocean acidification threaten communities across Oceania (Johnson and Vertessy 2014; Ahmed et al. 2011; Fabricius et al. 2013). These are but a few of many issues of current social and environmental concern to kam'āina, many of whom assert the need for approaches to natural resource management that can better accommodate local needs, interests, knowledge, and environmental variability than those being used in the islands today.

4.3   The *Ho'Ohanohano I Na Kupuna Puwalu Series* and Subsequent 'Aha

Overview

It was in the context of ongoing constraints on indigenous use and oversight of marine resources in Hawai'i, and with the objective of implementing a process to involve the Native Hawaiian community in the federal fishery management decision-making processs, that a series of 'aha (meetings) was first conceived by the Western Pacific Regional Fishery Management Council (here alternately termed "the regional fishery management council," "the fishery council," and "the Council"). Like the 'aha lā'au lapa'au that sought to preserve and advance traditional healing and medicine, and the 'aha 'olelo that sought to advocate use of the Hawaiian language in decades past, the 'aha mahele kumu-waiwai ho'onohonoho ku'una sought to advance traditional natural resource management for application in the islands now and in the years to come. In keeping with diverse needs and interests, the meetings involved broad representation and elicitation of varied perspectives on natural resource management strategies, along with in-depth deliberation on problems and constraints that confront Native Hawaiians and other kama'āina who seek to use and conserve island ecosystems and resources in traditional fashion. The meetings were inspired by the vision of John W. E. K. Ka'imikaua,, a well-known kumu hula and Hawaiian scholar, who had a vision, based on an ancient prophecy, that it was time for the Hawaiians to rise up and become prominent in the protection of island ecosystems and resources (DaMate 2016).

By way of background, the Western Pacific Regional Fishery Management Council is one of the eight regional fishery councils established by the Magnuson-Stevens Fishery Conservation and Management Act of 1976 to oversee marine fisheries conducted in the Exclusive Economic Zone (EEZ) of the United States and its territories. In the Western Pacific Region, the EEZ encompasses federal jurisdiction waters (three to 200 nautical miles) surrounding the entirety of the Hawaiian chain, American Samoa, Guam, the Commonwealth of the Northern Mariana Islands (CNMI), and various remote island areas. The Western Pacific Region is truly vast, encompassing nearly 1.5 million square nautical miles or 48% of the overall U.S. EEZ. The Council, supported by a staff of fishery specialists and administrators, is comprised of 13 voting members and three non-voting members, including: eight nominated by the governors of Hawaii, Guam, American Samoa and CNMI, and appointed by the Secretary of Commerce; four designated state officials; and four designated federal officials representing the National Marine Fisheries Service (voting) and the U.S. Department of State, U.S. Coast Guard, and U.S. Fish and Wildlife Service (non-voting). The Council and Council staff regularly interact with the U.S. Department of Commerce, NOAA Fisheries and with state, territorial, and commonwealth agencies to monitor regional fisheries and to develop and adjust science-based policies for ensuring their appropriate use and sustainability over time. The Council and Council staff also continually interact with regional fishery management organizations, such as the Western and Central Pacific Fisheries Commission and the Inter-American Tropical Tuna Commission to develop international conservation and management measures for highly migratory species such as tuna. Given ecological linkages among inshore, nearshore, and offshore biota and habitats, and routine use of multiple zones by island residents, interjurisdictional fisheries research, broad public participation, and challenging policy decisions are common in this unique region.

Recognizing the fundamental importance of marine fisheries to indigenous residents across the U.S. Pacific Islands, the Council plays an important role in developing and administering programs focused on "sustaining native fishing rights and the participation of the indigenous people of the region who depend on the sea to fulfill their nutritional and other needs" (Ka'ai'ai 2016: 2). As important as this mission is to indigenous societies in the region, it has not been easy to accomplish—in part because traditional-indigenous value systems emphasize collective needs and interests—a social fact that can be lost or ignored in decision-making processes that are heavily influenced by the majority cultural emphasis on the interests, rights, and property of the individual—as embraced by many visitors and residents of present-day urban Hawai'i, for instance. Ka'ai'ai (2016: 2) discusses this dilemma as an essential consideration in marine policy decisions that have the potential to impact indigenous and local society and culture across the U.S. Pacific Islands:

> Although communities at large were included in the new Council process [beginning in 1976], [macro-social] barriers prevented recognition of indigenous or native fishing rights and practices [by cognizant government agencies]. This is not uncommon as the United States, through various Congressional actions, has sought to protect cultural values of the native people while at the same time disinheriting the native people of their natural resources by allowing the privatization of some of them and exercising eminent domain to provide benefits to its citizens through the creation of public trusts and public domains out of native resources. The United States is not alone in its inability to protect traditional native people and communities. Nations in general have been poor custodians of native and traditional natural resource assets. Democracies have been successful in protecting individual rights but have not been successful

in protecting communal and traditional rights, particularly when they involve their own native people. These rights need to be protected to ensure survival of the native, traditional cultures and communities.

Sociopolitical and cultural challenges notwithstanding, the regional fishery management council recognizes the potential for traditional ecological knowledge and traditional forms of resource management to enhance marine fisheries across the region. This too is expressed by Ka'ai'ai (2016: 2), who asserts that participatory planning is a way for indigenous and local residents to contribute directly and substantially to the present-day natural resource management process:

> The traditional islander ways of managing and utilizing resources are empirical, time-tested methods whose success can be measured by the survival of the cultures that developed them. The wrong methods and practices did not survive. Today, the Western Pacific Regional Fishery Management Council process of public participatory decision-making can level the playing field for the resolution of conflicts related to marine resources use and management arising out of [historic] colonizing activities by the United States in Oceania. The Council process also provides an avenue for indigenous rights and knowledge to deliver benefits to the native communities and improve fisheries in the US Pacific Islands.

As is the case for the regional fishery management councils elsewhere in the United States, the Western Pacific council, in conjunction with NOAA Fisheries and state-level agencies, routinely engages in standard fishery science and management processes. The council system as a whole facilitates participatory input from scientists, fishery participants, and other knowledgeable persons involved in the use and/or scientific study and management of regional fisheries. According to the collective of regional fishery management councils, "fishery plans and specific management measures (such as fishing seasons, quotas, and closed areas) are developed based on sound scientific advice, and are initiated, evaluated, and ultimately adopted in a fully transparent and public process." For the Western Pacific and North Pacific councils, both of which manage fisheries involving numerous indigenous residents, this process is mandated by the Magnuson-Stevens Fishery Conservation and

Management Act to include input from native fishery participants and to incorporate local and traditional knowledge so as to enhance management of regional fishery resources (U.S. Department of Commerce, NOAA Fisheries 2007: 109).

Based on this mandate and rationale, and given the status of coral reef and nearshore fishery resources and the interests of kama'āina in advancing traditional knowledge and management approaches, Council staff members worked with Native Hawaiian organizations and state and federal agencies to convene the first in a series of traditional resource management meetings in August 2006. This and two subsequent meetings comprised what was termed the *Ho'ohanohano I Na Kupuna* (honor the ancestors) *Puwalu* (persons working together in unison) series. Co-convening groups and agencies included the Association of Hawaiian Civic Clubs, the Office of Hawaiian Affairs, the State of Hawai'i Office of Planning Coastal Zone Management Program, the Hawai'i Tourism Authority, Kamehameha Schools, and a large number of cultural practitioners and natural resource experts from communities around the island chain. The meetings were of unprecedented scope and scale in the modern era, involving extensive involvement and representation on the part of knowledgeable kama'āina from all the main islands and from many relevant fields of expertise.

In sum, the inital *Ho'ohanohano I Nā Kūpuna Puwalu* series, and 'aha that followed, enabled many scores of Native Hawaiian and other local representatives from moku on each of the main islands to discuss the prospective formalization of traditional ahupua'a- and moku-specific strategies for managing natural resources around the islands. Well-known public officials and highly respected kūpuna played particularly important roles, contributing long years of experience and knowledge to a deliberative process intended to answer a basic and overarching question: *How can traditional knowledge regarding use and management of natural resources be effectively incorporated into contemporary management processes and decisions undertaken by governing agencies in the Hawaiian Islands?*

The term 'aha has two related meanings in Hawaiian: meeting and standing council. The initial *Ho'ohanohano I Nā Kūpuna Puwalu* series resulted in the formation of two types of standing councils and related procedures. One is termed 'Aha Moku, comprised of representatives from the land districts on each of the main islands. Each ʻAha Moku island council interacts with representatives from the ahupuaʻa and moku on their island. A higher level council called the ʻAha Kiole Advisory Committee, comprised of a single representative from each island, was established by law in 2007 to guide the initial process of advancing a traditional place-based system of resource use and management (Act 212, Session Laws of Hawaiʻi 2007).

Significantly, the initial *Hoʻohanohano I Nā Kūpuna* Puwalu series was successful in contributing to the base of information and necessary impetus needed to develop legislation that gives indigenous and other local residents meaningful opportunities to advise government agencies on local and regional natural resource matters that are at the core of Hawaiian culture. The ʻAha Kiole Advisory Committee was eventually superseded by an entity called the ʻAha Moku Advisory Committee (AMAC), which by law is emplaced in the State of Hawaiʻi Department of Land and Natural Resources (Act 288, Session Laws of Hawaiʻi 2012). The nature of the *Hoʻohanohano I Nā Kūpuna Puwalu* series and subsequent ʻaha is summarized in the following pages.

Puwalu ʻEkahi: No Nā Lae'Ula
(Expert Practitioners of Tradition)

The first meeting of the *Hoʻohanohano I Nā Kūpuna Puwalu* series, held during August 15–17, 2006, was in large part an extensive and engaging set of introductory interactions between Native Hawaiians hailing from islands across the entire inhabited chain. Local representatives of other ancestries also attended the multi-day meeting, in keeping with the mixed demographic nature of the typical community in Hawaiʻi. The ʻaha included representatives of the relatively small and minimally populated islands of Niʻihau (pop. 170 persons in 2010) where the vast majority of residents are first language speakers of Hawaiian, and Lānaʻi (pop. 3102 in 2010), where a plantation-based economy has recently shifted primarily to tourism. Uninhabited Kahoʻolawe was also represented—by members of the Kahoʻolawe Island Reserve Commission.

On the order of 200 persons attended the initial meeting, including many kūpuna, cultural practitioners, educators, and activists well known in the Native Hawaiian community. Although many of those present had met on prior occasions, many had not, and meeting-specific introductions tended to be organized by island, with Maui participants, for example, presenting as a coherent social group to representatives from other islands, some of whom presented in a similarly organized way. Introductions tended to involve discussion of one's lineage and tenure in a given place, along with mention of social-environmental problems then occurring in certain island areas, most of which involved intrusion and misuse of natural resources by newly arriving residents. Kamaʻāina from around the island chain asserted their deep attachment to island and sea, and to lawaiʻa, mahiʻai ana, and other important traditional and cultural practices that have long been undertaken across the various mokupuni, moku, and ahupuaʻa.

Given fears that local-traditional knowledge could be exploited and result in further depletion of natural resources in a given ahupuaʻa, many participants were reluctant to share detailed knowledge of the physical environment, fishing or farming techniques, or other kinds of ʻike (knowledge) known to be used in the past and present. This reticence relates in part to respect for kūpuna who maintain the centuries-long tradition of safeguarding natural resources for use and oversight by resident ʻohana and in part to centuries of failed trust in outsiders. It was made patently clear by many Native Hawaiians attending the meeting that although governing agencies now claim ownership of the islands and assert kuleana (responsibilities) for managing natural resources, fundamental issues remain unresolved. That is, the overthrow of the Hawaiian monarchy is seen by many as illegal and invalid, statehood is not universally recognized, and many consider themselves to be the *actual* stewards of land and sea. Therefore, in some or many instances, government mandates and bureaucracy are seen as artificial layers covering the true rights and duties of Kanaka Maoli to care for the islands. Resistance continues.

Virtually all participants agreed that traditional ecological knowledge does indeed belong to the kūpuna and that no one should be able to use such knowledge for profit. In this respect, there was extensive discussion regarding the documentation of traditional resource use practices. Many participants asserted that if resource use traditions and traditional ecological knowledge are not documented, these cannot be adequately protected and may eventually be lost to time. Although it was asserted that formal-written documentation is not a traditional indigenous practice, its present value as means for clarifying the importance that natural resources have had to Native Hawaiians in recent centuries was also argued. In the end, participants concurred not only that palapala (written documentation) could help protect natural resources in ahupua'a and moku around Hawai'i, but also that great care must be taken in its distribution and use.

Perspectives on certain issues broached during the meeting varied extensively, and discussion was often impassioned—as might be expected of an initial meeting of representatives from so many island areas, each with its own history, socioeconomic challenges, and social-ecological conditions. But consensus was ultimately reached on many topics, including the following:

- The importance of cultural protocol and the need to treat traditional knowledge itself with hō'ihi (respect) both during the meetings and in general since this is such an important element of Native Hawaiian heritage;
- The need for residents and representatives from the various ahupua'a and moku to be able to undertake place-based resource management in different ways per varying ecological conditions and traditional resource use and management practices;
- The spiritual significance of the moon and the practical guidance provided by the Hawaiian lunar calendar, which indicates both the appropriate timing of actions such as planting certain crops and fishing for certain species, and limitations on these actions, such as prohibitions on harvesting certain species of fish at certain times of the year to enhance spawning and population growth potential;
- The need to effectively and consistently communicate cultural knowledge and practical natural resource experience across generations of Native Hawaiians, and to ensure that outsiders understand the basic tenets of traditional ahupua'a based management of local resources;
- The cultural-dietary importance of limu (edible seaweeds), dissatisfaction with the way newly arriving ethnic groups were (at the time)

overusing limu and misusing the reef and shoreline ecosystems that nurture limu, and the exemplary value of a Native Hawaiian limu growout project on Molokaʻi;

- The need to preserve Native Hawaiian fishing and spiritual practices and associated values for future generations, and the challenges of doing so without disseminating information that could lead to misuse of ahupuaʻa-specific natural resources;
- The need to work on addressing marine pollution and problems with coral reef ecosystems around the islands, including coral bleaching events associated with warming oceans;
- The need to address problems with invasive species and their impacts on native biota and traditional use of native and culturally significant plants and animals;
- The need to examine the efficacy of agency-based enforcement of existing natural resource use regulations, and the role that local customs and social sanctions might play in reducing natural resource problems that occur in various ahupuaʻa despite modern government-based regulatory processes;
- The need to address the marine-environmental impacts of coastal development, including: the effects of runoff from roads, golf courses, resorts, and other areas where human presence is extensive; siltation of coral reefs; and overfishing;
- The need to address the damaging effects that newly immigrating people often exert on marine resources in the absence of local-traditional knowledge;
- The need to advance the preservation, renovation, and contemporary use of ancient fish ponds;
- The need to examine and regulate the use of motorized vehicles on Native Hawaiian lands, and the effects of tour boats and jet skis on marine ecosystems;
- The need for thorough assessment of impacts resulting from issuance of land use permits on Native Hawaiian lands; and
- The need to amend overarching problems related to loss of native fishing rights and resources in Hawaiʻi by formally reintroducing historically effective ahupuaʻa-based resource use and management practices.

Inset B: Senator Daniel Akaka, Keyonote Speaker at *No Nā Lae'ula*

U.S. Senator Daniel Akaka was keynote speaker at the initial meeting of the *Ho'ohanohano I Nā Kūpuna Puwalu* series. Mr. Akaka was America's first senator of Native Hawaiian ancestry. Raised in Pauoa Valley on Oʻahu, he attended Kamehameha Schools, earned a Master of Education degree from the University of Hawaiʻi at Mānoa, and worked as a teacher and principal in the Hawaiʻi school system for 18 years. He was elected to the U.S. Senate in 1990, re-elected in 1994, 2000, and 2006, and throughout his tenure worked consistently to advance the interests of Native Hawaiians, other Pacific Islanders, and other indigenous groups in the United States. In 2000, the senator introduced the Native Hawaiian Government Reorganization Act, which eventually became known as the Akaka Bill. The bill evolved to address a wide range of Native Hawaiian interests, many legal complexities, and ongoing resistance in the Senate. As noted by Lindsey (2002: 1), a fundamental paradox constrained passage of the bill: "a government-to-government relationship with the United States will benefit the Native Hawaiian peoples [the intent of the bill]...[but]...the creation of a government-to-government relationship will not settle the Native Hawaiian peoples' international claims against the United States."

Senator Akaka welcomed attendees of the first 'aha in his inimitably humble and gracious manner and stated his deep interest in safeguarding traditional Hawaiian practices through public policy. He respectfully acknowledged kūpuna past and present and their intimate knowledge of natural resources, ocean currents and weather conditions, and traditional ways of farming, fishing, and navigation. The senator asserted that traditional resource management practices are needed now more than ever and that by incorporating well-tested knowledge into the modern regulatory process, natural resources and Native Hawaiian culture would only benefit. It was his perspective that the younger generations must now be encouraged to assume responsibility for future stewardship of Hawaii's natural resources.

Daniel Kahikina Akaka retired from the Senate at the age of 88 and passed away on Oʻahu on April 6, 2018.

Discussion of certain topics was intense and protracted. For example, participants deliberated extensively on the complex process of ensuring that government agencies in Hawaiʻi work to protect the rights of Native Hawaiians and the natural resources that sustain indigenous people and culture. Many present voiced dissatisfaction with the trend toward establishment of no-harvest marine protected areas in the islands, and the implications that permanent closure of fishing areas might have for Native Hawaiian fishermen, their families, and the communities of which they are a part. It was stated clearly that permanent closure of fishing grounds is not a traditional resource management concept.

The then-recent ban on 'upena'apo'apo (gill nets) established by the State was also discussed at length. Some participants asserted that the ban unfairly punishes many for the actions of a few indiscriminate fishermen who leave nets unattended or who overfish certain areas. Such behavior was said to be in contrast with more preceise and careful traditional use of nets, as overseen by the konohiki (local supervisor of the ahupua'a). Participants proposed that expanded monitoring of fishing activities was needed and that the konohiki and kapu-based systems of management would help reduce misuse of nets and other gear.

It was reiterated that the konohiki system involves direct observation and monitoring of resources and fishing practices by a locally based cultural expert, and that the kapu system involves periodic prohibitions and restrictions on certain nearshore and offshore fishing and shoreline gathering practices. While both of these traditional forms of resource management were undertaken in different ways and times and places in different moku and ahupua'a over time, it was emphasized that the basic principle of expert local oversight was the both historic norm and essential for effective management in contemporary Hawai'i. Notably, as discussed in Chapter 3, in many areas, ahupua'a boundaries—Keawa'ula on O'ahu, for example—incorporated the open ocean, and thus the konohiki oversaw and could regulate pursuit of pelagic resources from the shoreline well out to sea (Kosaki 1954; Meller 1985).

Participants in the meeting universally called for reestablishment of traditional ahupua'a-based management of natural resources. Because many Native Hawaiians and other kāma'āina retain and refine extensive knowledge of marine and terrestrial resources in and around their respective areas of residence, or those of their larger 'ohana, there was strong assertion that functioning ahupua'a are both possible and highly desired. Concerns about limitations on fishing and gathering activities that might be established using the traditional ahupua'a approach were stated by a few meeting participants, but it was made clear by others that such strictures were and are intended to conserve and enhance the resource base and that there would be ample opportunities, as in days of old, for residents to pursue marine and shoreline resources both within and across ahupuaʻa boundaries—provided that they comply with prospective local rules, customs, and cultural protocol. Such customs include, for example, the requesting of permission to gather or otherwise utilize resources in other ahupuaʻa. Development of the kanāwai (rules) for using and managing marine resources in and across given ahupuaʻa would be the responsibility of local- (ahupuaʻa) and district-level (moku) councils, with the specifics of how this would occur discussed in subsequent pages of this text. In all cases, rules would be developed through consultation with kupuna and other key representatives of local communities.

In the final hours of the ʻaha, participants drafted a resolution to "begin the process to uphold and continue Hawaiian traditional land and ocean practices into the governance and education of the Hawaiian archipelago." The resolution, approved and adopted on August 17, 2006, is provided in its final form below (Fig. 4.2).

Fig. 4.2 A group of photo-ready participants at No Nā Laeʻula, the first meeting of the Puwalu series

Resolution to unite Native Hawaiians to move forward, to live, to grow, to gather together, to stand firm and to restore and perpetuate the Hawaiian way of life.

WHEREAS, more than 100 elders, parents and youth—who are traditionalists, practitioners and experts as well as lineal descendants of the original inhabitants of the islands, Kure Atoll, Midway Atoll, Pearl and Hermes Atoll, Lisianski Island, Laysan Island, Maro Reef, Gardner Pinnacles, French Frigate Shoals, Necker Island, Nihoa, Niʻihau, Kauaʻi, Oʻahu, Molokaʻi, Lānaʻi, Maui and Hawaiʻi - met to honor our ancestors in the first of a series of conferences;

WHEREAS, this first conference provided distinguished elders, practitioners and experts a forum to discuss and share the cultural practices of the fishermen and the farmers from the ahupuaʻa of 45 traditional land districts of the Hawaii archipelago;

WHEREAS, the participants acknowledged that the spiritual and physical well being of indigenous people of Hawaii are intrinsically tied to the land and the sea;

WHEREAS, the participants recognized that the knowledge they share and hold reflects thousands of years of experience sustaining the resources of the land and the sea;

WHEREAS, the participants identified examples of impacts negatively affecting their access to, and the abundance and availability of, the natural resources;

WHEREAS, the participants reaffirmed to move forward together with one voice as lineal descendents and urge the Hawaiian people and supporters of Hawaiian culture to rise up to ensure the community's health, safety and welfare;

NOW, THEREFORE, BE IT RESOLVED that those attending this conference call on Hawaiian people to begin the process to uphold and continue traditional Hawaiian land and ocean practices in the governance and education of the Hawaiʻi archipelago;

NOW, THEREFORE, BE IT FURTHER RESOLVED that the conference participants call for perpetuation and preservation of the knowledge of practitioners and the restoration of healthy ecosystems through furtherance of the ahupuaʻa management system, konohiki management, kapu, hoaʻāina rights and the re-establishment of ʻAha Moku.

Finished is the stealing of the land; finished is the stealing of the sea; finished is the stealing of the life of the land. The people of the land shall rise up.

- Wanana prophecy

Puwalu 'Elua: Ke Kumu 'Ike Hawai'i
(Source of Hawaiian Knowledge)

The second Puwalu, *Ke Kumu 'Ike Hawai'i*, held November 8–9, 2006, was attended by well over 150 educators, cultural practitioners, and natural resource experts from around the islands and island districts. The 'aha focused on indigenous ways of knowing natural resources and ecosystems, and the prospective role of such knowledge systems in Hawaii's educational curricula. The intent was to examine educational initiatives that could promote awareness of ahupua'a-based cultural practices and integrate traditional knowledge into the standard school curriculum. The educational objectives of the meeting, as co-convened by the regional fishery management council, also addressed provisions of Section 109-479 of the Magnuson-Stevens Fishery Conservation and Management Act, which calls for marine education and training programs that can foster practical knowledge and technical expertise regarding stewardship of living marine resources among native residents of the Western Pacific (and Alaska).

Educators present at the conference included Hawaiian language immersion specialists, charter school delegates, persons working in private school settings, and representatives from the State of Hawai'i Department of Education. Participants discussed challenges involved in the prospective reestablishment of ahupua'a-based natural resource management, and it was ultimately agreed that education and outreach would be an indispenable part of the solution.

During the meeting, Native Hawaiian cultural practitioners provided educators with insight into the traditional values and practices discussed during the initial Puwalu of the series. There was much emphasis on the desirability of indigenous Hawaiians continuing the tradition of sharing knowledge between kūpuna and keiki, and it was agreed that teachers could and should support 'ohana in the transmission of knowledge between generations, and in educating children about proper care and use of the natural world around them. There was also discussion about the need for Hawaiian teachers to educate children who do not have kūpuna from whom they might otherwise learn about traditional treatment of the natural environment.

Educators discussed ways to incorporate traditional knowledge into pilot curricula that could be tested and eventually used in schools throughout the islands. Some participants suggested that opportunities for experiential learning should be developed and that lessons should be moku-specific. But, because many teachers in the educational system are not Native Hawaiian, a cultural practices training program would be essential for initiating such a process. It was determined that any future hands-on learning program would require the acceptance and guidance of knowledgeable individuals, families, and hui (organizations) throughout the islands.

The educators also discussed potential obstacles to teaching traditional knowledge and practices in the classroom. Many felt that, although the effort would be highly rewarding, incorporation of place-specific traditional knowledge into lesson plans would likely involve administrative and practical challenges, including:

- How to categorize and manage an inventory of diverse cultural practices;
- How to decide what is most essential to teach, and how to teach in a way that is age-appropriate;
- How to standardize the terms used to communicate a curriculum involving traditional knowledge;
- How to devise a holistic approach to teaching that incorporates the needs of parents, neighbors, and the larger community;
- How to incorporate traditional knowledge and experiential learning into a system that otherwise emphasizes Euro-American topics and perspectives; and
- How to measure that actual attainment of traditional knowledge.

The educators identified existing educational policies that have the potential to hinder incorporation of traditional knowledge into contemporary curricula around the state. For instance, some participants anticipated difficulty gaining permission to bring kūpuna into the classroom as instructors. Permission for off-campus field trips, essential for

Fig. 4.3 Makua kāne (father) and kaikamahine (daughters) ready to throw net, South Kona, 1997

hands-on learning, was also thought likely to involve administrative hurdles (Fig. 4.3).

Participants considered a range of prospective solutions to such problems. For instance, it was suggested that a policy statement could be obtained from the Department of Education or State Legislature affirming the importance of teaching Native Hawaiian history and culture in the classroom. Educators also considered ways and means for enabling cultural practitioners to be allowed to teach. This could involve an accreditation system based in part on years of hands-on experience and accumulated wisdom rather than years of formal schooling. Finally, educators stressed that if Native Hawaiian studies are not to be part of the core curriculum, students should at least be given the opportunity to undertake such study.

The second ʻaha of the series generated observably heartfelt statements of positive affirmation about the future. Participants agreed to (a) apply what they had learned to development of strategies for incorporating traditional and customary knowledge into school curricula, (b) seek out and learn from valid sources of traditional knowledge and culture such as prominent kūpuna, (c) establish and maintain rapport with cultural practitioners and educators in their communities, and (d) create a website and/or mail list to advance the sharing of traditional knowledge and culture, ideas for new curricula, and lesson plans.

While Ke Kūmu 'Ike Hawai'i was focused primarily on formal and informal modes of education and the content of contemporary and future educational plans in the islands, there was also more general, ongoing discussion of natural resource issues and a system through which Native Hawaiians could contribute directly to natural resource decision-making across the islands. Participants made the following recommendations in this regard:

- 'Aha Moku (standing district-level councils) that are responsive to needs and interests of the respective ahupua'a in a given district (as potentially determined by prospective 'Aha Ahupua'a) should be established on each island [it was acknowledged the some ahupua'a are not at present sufficiently populated or organized to develop a standing council and that people in certain other ahupua'a around the islands have long been organized to address natural resource issues];
- Laws should be developed to prohibit introduction of invasive species and to remove alien species that are already affecting the environment;
- An inventory of natural resources of importance to Native Hawaiians and other kāma'āaina should be conducted across the islands, and a monitoring plan should be established to gauge changes in such resources and associated ecosystems;
- State and county governments should establish means for community-based enforcement of rules and practices associated with use of natural resources in each ahupua'a; and
- A stateholiday should be established to celebrate and honor Kānaka Maoli and their deep knowledge of island ecosystems and resources.

Inset C: Key points made by Mr. Colin Kippen, J.D., keynote speaker at _Ke Kumu 'Ike Hawai'i_

Mr. Colin Kippen, M.A., J.D., addressed a variety of issues during the second meeting of the _Hoohanohano I Na Kupuna_ Puwalu series. Mr. Kippen is former Executive Director of the Native Hawaiian Education Council, Chair of the U.S. Secretary of the Interior Review Committee of the Native American Graves Protection and Repatriation Act, and Coordinator for the State of Hawaii's homeless program under the Abercrombie administration. He is an experienced Native Hawaiian leader on indigenous rights issues related to education, poverty, homelessness, environment, environmental justice, and self-determination.

Mr. Colin Kippen. Mr. Kippen states that while he is not a traditional practitioner nor is he reliant on subsistence practices for survival, he _is_ a Native Hawaiian who dedicates much of his time working to preserve and protect indigenous rights and practices. Native Hawaiians, he explains, have the birth-given right to interact with a given natural resource in a given place. He states that, for Hawaiians, a native right is akin to kuleana (responsibility): "if you have kuleana, what that means is you have a practice, and if you have a practice, then you have traditions and customs that will live on." But he acknowledges the difficulty of developing a singular approach that can protect all such practices since these tend to differ by moku. He asserts that government entities must therefore work to accommodate all the between-moku differences and conditions, and Native Hawaiians must come together in order to have a say in how these resources are managed. By way of example, Mr. Kippen recounts a lengthy debate in which one faction of a Native Hawaiian organization was opting to support the state's then nascent ban on gill netting, while another rejected the ban and instead asserted the need for traditional Native Hawaiian management practices that could better guide the use of nets. Mr. Kippen states his belief that those in favor of endorsing the state's ban did not fully understand the need for native kuleana of the resource and instead put their trust in external agencies and regulations. With much experience in this and related matters, he states that this kind of situation is not uncommon, noting that "somehow native people always get tangled up [between tradition and the agencies and exigencies of the present-day]—we end up in a position where we have to defend our rights even to access the resource and protect and practice our traditional ways." Given this situation, he says that "if Hawaiians can document the actual nature of resource use practice in a given area, we can ultimately regulate and enforce it as well." He says that documentation makes people nervous but stresses its importance. "If a practice is not documented," he explains, "it cannot be protected." Documentation is a way to establish and claim kuleana in a context of changing times and circumstances and can help Kanaka Maoli to understand their kuleana. But

he stresses that government agencies have to honor the rights and capac-
ities of Hawaiians to manage resources of land and sea. He asserts that
the primary issue at hand is how to create an ahupua'a-based system of
management. "This will be a politically difficult but not impossible task,"
he says, and "education is the key to success." He uses the example of
Nā Lau Lama Education Initiative, which incorporates aspects of Hawaiian
culture into the state's public school system, in which some 60,000 Native
Hawaiian students are now enrolled. Mr. Kippen stresses the importance
of such programs in better communicating Native Hawaiian values and
cultural practices to government decision-makers, educators, and the
public at-large.

Puwalu 'Ekolu: Lawena Aupuni (Policymakers)

The third Puwalu of the series focused on development of policy options
for reestablishing ahupua'a-based resource use and management practices.
Many challenges were identified in this regard, and participants repeatedly discussed the need for policy-makers to address the many competing interests that characterize use and management of natural resources in
contemporary Hawai'i. These include the needs of the indigenous versus
those of society at-large, use versus preservation of natural resources, and
informal modes of governance versus formalized laws and policies.

Contemporary challenges and competing interests notwithstanding,
participants universally called for reestablishment of traditional resource
use and management strategies. Discussion repeatedly returned to the fact
that Native Hawaiians had, over many centuries, developed highly effective approaches to care for and use resources of island and ocean. There
was also much discussion about the changes and associated challenges
that had occurred following the arrival of new people and ideas in the
islands, and the need to recognize the value of the ecological knowledge
that has been accumulated and is still being developed and used by Native
Hawaiians today. This sentiment was captured in the opening remarks by
Ms. Kitty Simonds, Executive Director of the regional fishery council:

> I welcome you today as agents of change in Hawai'i, ready to shift and
> advance the way we view and manage our natural resources…we are not

blind to the signs that foretell destruction of our natural resources and our native culture unless something is done now. Our shift into the future is a step back to retrieve and revive the native culture of Hawai'i that was supplanted by Western culture. It is a long overdue step to recognize the value of the culture that existed for millennia in these islands and which is embodied in the cultural practitioners who are gathered with us today. This valuable inheritance is available for all of us if we are willing to accept it.

Much of the meeting involved discussion of the process through which the interests, values, needs, and knowledge of the indigenous people of Hawai'i could be formally incorporated into government decision-making. In indigenous terms, this would involve the establishment of 'aha to facilitate local and regional deliberation and decision-making, with oversight by an overarching 'aha that could readily interact with extant agencies involved in management of Hawai'i's natural resources. It was determined that this general process could be configured differently for different islands, but with the overarching strategy that persons specializing in ahupua'a-based fishing and agricultural practices would help represent indigenous needs and interests in government decisions affecting the region's ecosystems and natural resource-dependent people.

At the time of the *Ho'ohanohano I Na Kupuna Puwalu* series, certain institutions had then recently initiated work to incorporate attention to traditional ecological knowledge and practices in regional planning documents. For instance, the 2006 *Hawai'i Ocean Resources Management Plan* (ORMP) sought to address these issues. 'Aha speaker Mr. Doug Tom, then Program Manager for the Hawaii Coastal Zone Management Program, stated that the new plan would be founded in part on three Native Hawaiian perspectives: the interrelatedness of the land and the sea; preservation of ocean heritage; and promotion of collaborative governance and stewardship. "Resource management," he pointed out, "is really more about managing human activities that affect the resources than managing the actual resources." Mr. Tom asserted the importance of partnerships between government agencies, private industry, and communities, and noted that a moku-based management

framework may act as a common thread in fulfilling coastal zone mandates and activities across the islands. Subsequently updated in 2013,
the ORMP describes the progression of the 'Aha Moku process, the
legislation that ultimately gave it life (described further along in this
chapter), and varying perspectives regarding the need for a formalized
approach to traditional resource management:

> During the Public Listening Sessions (PLS) for this ORMP Update
> [2013], community members gave input on what modern day ahu-
> pua'a management means to them. An ahupua'a encompasses a "slice"
> of land from the mountains to the sea, and the Native Hawaiian view
> is that the entire land division is integrated. People spoke of kumuwai,
> which means both the source of wealth as well as the source of a stream,
> and in this instance the source comes from the rain above to the tip of
> the mountain, traveling through the ahupua'a as a stream to the ocean.
> There is a reverence and acknowledgement that all is connected and that
> it is a higher power's will that brings all water from its starting point in
> the heavens above to the ocean that surrounds the islands. Managing an
> ahupua'a, while similar to the term conservation, incorporates sustaina-
> bility principles. The lo'i (irrigated terrace) that feeds poi to the people of
> an ahupua'a also functions as a place where non-point source sedimen-
> tation occurs, slowing down the flow of water so that it can recharge the
> water table below the soil. Community members expressed a need to feed
> their community, especially the kūpuna (elders) and keiki (children) who
> were unable to catch or grow food themselves. They saw traditional ahu-
> pua'a land management as a way to ensure food for now and to sustain
> it for the future. A community working together can plant and maintain
> lo'i, reconstruct their shoreline Native Hawaiian fishponds, gather their
> own pa'akai (salt), and keep their stream inflows to ensure a recharg-
> ing of water in the entire water cycle. Many felt that they could do this
> without waiting for government assistance and without a statewide plan
> to tell them how to manage their own land. Others were organized for
> their entire island's natural resources, such as on Moloka'i, and wanted
> [an] 'Aha Moku system to be mandatory. (State of Hawaii, Office of
> Planning 2013: 68)

Some participants in the Puwalu series stated that various state agencies were not at that time adequately addressing the needs of Native

Hawaiians in the cultural assessment process. It was asserted that in order to make environmental assessment (EA) and environmental impact assessment (EIS) processes relevant for twenty-first century Hawai'i, existing laws and policies may need to be changed to better assess and accommodate the public interest. More specifically, it was felt that questions put forth during EA and EIS public comment periods often go unanswered or unaddressed, and that in Hawai'i, the process should be more sincerely responsive to the needs of Native Hawaiians. One solution discussed at the meeting was that at least one cultural practitioner involved in the 'Aha Moku process should have direct input into any and all resource state and federal EA- or EIS-related decisions that could affect indigenous culture, heritage, customs, traditions, or resources.

Numerous participants also discussed concerns about the enforcement dimension of natural resource management in Hawai'i. The prominent sentiment expressed at the meeting was no only that reinstatement of ahupua'a-based resource management would likely improve protection of marine and terrestrial resources, but also that the current level of regulatory enforcement may well be inadequate to address locally relevant kanāwai. This was thought to be most likely in the more remote moku and ahupua'a. For this reason, a variety of prospective means for ahupua'a residents to assist in or enable local enforcement were discussed in terms of consistency with the konohiki approach and how it could be coordinated with extant state and federal policies, measures, and personnel.

The potential benefits of the ahupua'a system were reviewed by the Puwalu. A participant from Kaua'i discussed the need to address disasters and the need for response planning in rural parts of Hawai'i. From his perspective, first response to Hurricanes 'Iwa in 1982 and 'Iniki in 1992 was weak in remote areas of hard-hit Kaua'i, and that disaster-prepared local residents could have contributed more effectively. The participant asserted that the ahupu'a and moku-based system of representation and-communication is ideal for organized disaster response.

Participants also discussed the beneits of a locally organized system of natural resource management when responding to marine-environmental problems, such human-monk seal interactions or deposition of marine debris along the shoreline. One participant discussed olfactory and other problems related to a dead whale found on the shoreline of an ahupuaʻa on a neighbor island, her dissatisfaction with government agencies who might have responded to the situation more effectively, and the potential real-time communications role that a recognized konohiki might have played to alleviate the situation. In total, it was determined that effectiveness of response to localized problems would be enhanced by returning a measure of authority to konohiki, and by ensuring the representation and communication of local needs, concerns, and interests to extant government agencies through the ahupuaʻa and moku councils.

Inset D: Welcoming Remarks by Ms. Kitty M. Simonds, at *Lawena Aupuni*

Ms. Kitty Muller Simonds is a Native Hawaiian from the island of Maui who has served as the executive director of the Western Pacific Regional Fishery Management Council since 1983. She joined the Council in its early months after serving on the staff of U.S. Senator Hiram L. Fong in his Washington, DC, and Honolulu offices. On her watch, the Council is successfully committed to innovative marine resource management for federally managed fisheries of Hawaiʻi, American Samoa, Guam, the Northern Mariana Islands and the U.S. Pacific remote islands. For example, the Council's Coral Reef Fishery Ecosystem Plan was one of the nation's first eosystem-based fishery management efforts. In parallel with this initiative, the Council convened a series of workshops to capitalize on local and traditional ecological knowledge that could be adapted into a modern fishery ecosystem management framework. On the national level, she co-organized the First Stewards Symposia held in 2012 and 2014, which brought together the indigenous peoples and treaty- and non-treaty tribes of the United States to Washington, DC, to discuss the impacts of climate change upon them and the resources they can provide to help address climate change impacts on the nation.

Ms. Kitty M. Simonds. Ms. Simonds welcomes the participants and notes that many present have been working for years to materialize needed changes to the natural resource management paradigm across the islands. She comments on the work of the Association of Hawaiian Civic Clubs, the Office of Hawaiian Affairs, the State of Hawaiʻi Office of Planning, and the Hawaiʻi 2050 Sustainability Task Force in this regard, and points out that at statewide community planning meetings in 2006, the public strongly advocated the ahupuʻa concept and ecosystem stability as essential for ensuring resource sustainability in the years to come. Ms. Simonds summarizes the responsibilities of the Western Pacific Regional Fishery Management Council in working toward resource sustainability, and its accomplishments in establishing fishery ecosystem plans that account for the many human and biophysical factors and linkages that affect marine resources in the Pacific island region. She describes Council mandates to work closely with communities to develop appropriate policies that function to both conserve marine resources and provide for their continued use, and notes that the reauthorized Magnuson-Stevens Act directs the Council to "develop means by which local and traditional knowledge can enhance science-based management of fishery resources of the region"—an important objective of the Puwalu series. Ms. Simonds concludes by stating that it is clear that a critical mass has been reached to initiate a paradigm shift in the way the people of Hawai'i view, relate to, and manage natural resources; that the time is here, today, now, to establish a consultation process that will allow communities to develop plans for ensuring sustainability of resources within their ahupua'a. She says it is a great honor that Native Hawaiian cultural practitioners are willing to share the truths conveyed to them from their ancestors; that these truths cannot be lost but must flourish and be adhered to so that future generations of Hawaiians can benefit. Together, she says in closing, the way can be found to protect Hawai'i's cultural legacy and sustain Native Hawaiians and their island ways for the benefit of all of Hawai'i.

An important objective of the Puwalu series was to identify ways in which existing government and community programs and initiatives could be enhanced through reestablishment of an ahupua'a- or moku-based approach to the use and care of natural resources around the islands. Such programs and initiatives include several community partnerships established by the State of Hawai'i Department of Land and Natural Resources, including (a) a curator agreement with the Royal Order of Kamehameha I; (b) work with Kailua Hawaiian Civic Club to care for Kawainui Marsh on O'ahu; (c) support for 'Ahahui Mālama I Ka Lōkahi in its efforts to care for Ulupō Heiau; (d) an agreement with Pu'u Olai Wetland Management Association to enhance protection of wetland areas near Mākena on Maui, and (e) an agreement with the Hawaiian Civic Club of Wahiawā to care for the Kūkaniloko Birthing Stones. The Department of Land and Natural Resources also continues to assist local fishery management at Hā'ena on Kaua'i and other locations, and to implement the Mauka/Makai Watch Program, which facilitates local monitoring of resources of land and sea in eight communities around the islands. Agency representatives present at the meeting reported an eagerness to expand such partnerships to include prospective collaboration with community representatives attending the ʻaha, though limited interaction of this nature has occurred to date.

An important topic of discussion undertaken during the third meeting in the series addressed the nature of "Western" or Euro-American societal perspectives on resource management, as distinct from the perspectives typically held by Native Hawaiian cultural practitioners. As summarized in the Table 4.1, the perspectives were indeed envisioned by participants as being quite different, with implications for resource management in Hawaiʻi should the Native Hawaiian perspective be better incorporated into the existing systems of governance.

It was stated at the meeting that modern science and contemporary management approaches often do not effectively address whole systems or relationships between the human and biological components that comprise the ecological whole, but that systems of traditional knowledge do tend to be holistic in nature. As such, a traditional systems approach (that of the ahupuaʻa and moku), was said to be relatively more adaptable to changing real-time conditions and situations since it includes understanding of specific areas, and the nature of linkages among given areas. This was seen as particularly important given extensive variability and ongoing change among sociocultural, economic and biophysical conditions within and across islands, districts, and even ahupuaʻa. Participants asserted that differences in perspectives

Table 4.1 Competing perspectives on marine resource management

Topic	Euro-American perspective	Native Hawaiian perspective
Predominant purpose for using living marine resources	Commerce, recreation, consumption	Cultural traditions, consumption and sharing, recreation, commerce
Years of fisheries data collection	<100	>1000
Relation to the land	Ownership	Stewardship
Normative fishing ethic	Catch what you can	Catch what you need
Management horizons	Present and future (few generations)	Past, present, future (many generations)
Rules and regulations	Rigid	Adaptive

and experiences are not conceptual constructs—the indigenous people of Hawai'i have been developing an empirical, place-specific and general understanding of the land and sea for nearly two millennia. Native Hawaiian fishermen, farmers, and cultural practitioners assert that any valid approach to caring for natural resources must draw on knowledge and experience developed over this long course of history.

State of Hawai'i Act 212

As a result of efforts undertaken by key individuals attending the *Ho'ohanohano I Nā Kūpuna Puwalu* series, State Act 212 was passed during the 2007 Hawai'i Legislative Session. State Representative Mele Carroll (representing Kaho'olawe, Molokini, Lāna'i, Moloka'i, Keanae, Wailua, Nahiku, and Hana), and State Senator Kalani English (representing Hana, East and Upcountry Maui, Moloka'i, Lāna'i and Kaho'olawe), played particularly important roles in moving the Act through the legislature. The Act specifies creation of "a system of best practices that is based upon the indigenous resource management practices of moku (regional) boundaries, which acknowledges the natural contours of land, the specific resources located within those areas, and the methodology necessary to sustain resources and the community." It also establishes a temporary Governor-appointed eight member 'Aha Kiole Advisory Committee that would participate in meetings throughout the islands to aid in: (a) developing consensus on establishing an administrative structure for the 'Aha Moku system, to include a permanent 'Aha Kiole, the eight members of which will each represent one of the main islands; (b) a selection process and eligibility criteria for members and an executive director; (c) 'Aha Moku goals, objectives and benchmarks; and (d) an operational budget necessary to support the objectives and functions of the 'Aha Kiole and 'Aha Moku.

[State Act 039, passed in April 2009, extended the functioning of the temporary 'Aha Kiole into 2011. As noted previously and later in this text, the 'Aha Kiole Advisory Committee was ultimately superceded by an entity called the 'Aha Moku Advisory Committee or AMAC].

Subsequent 'Aha

Kūkulu Ka 'Upena (Building the Net)

Following the initial *Ho'ohanohano I Nā Kūpuna* Puwalu series, an additional conference was convened during April 10–11, 2007, to enable the conceptual structuring of the standing councils. The meeting was termed *Kūkulu Ka 'Upena* (Constructing the Net), wherein the 'upena (fishing net) represents the capacity of Hawaiians to interact productively under an interwoven social system that enables both unity in the vivifying of tradition (the interwoven net), and representation of place-specific conditions, needs, knowledge, and interests (the individual strands). Like the previous meetings, the 'aha was well-attended (Fig. 4.4).

The meeting focused especially on the basic criteria needed for choosing standing representatives and identification of the most salient issues that continue to affect Native Hawaiians in each of the 37 moku across the islands. Historic government maps depicting ahupua'a boundaries were updated with previously undocumented boundaries that are commonly known to Hawaiian practitioners and their 'ohana but that had never been formally recorded. An example of the nature and extent of

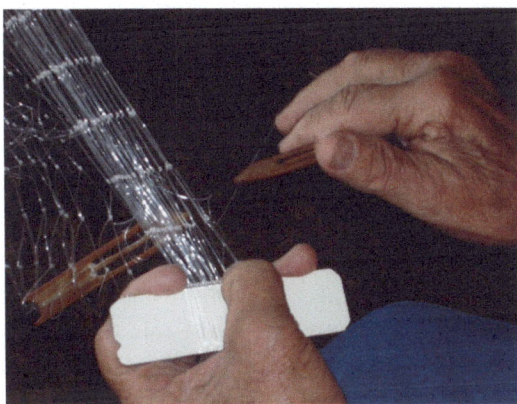

Fig. 4.4 The creation of an 'upena (fish net) symbolized Puwalu 'Eha, which reconvened traditional practitioners to structure an 'Aha Moku council system (Photo courtesy of the Western Pacific Regional Fishery Management Council)

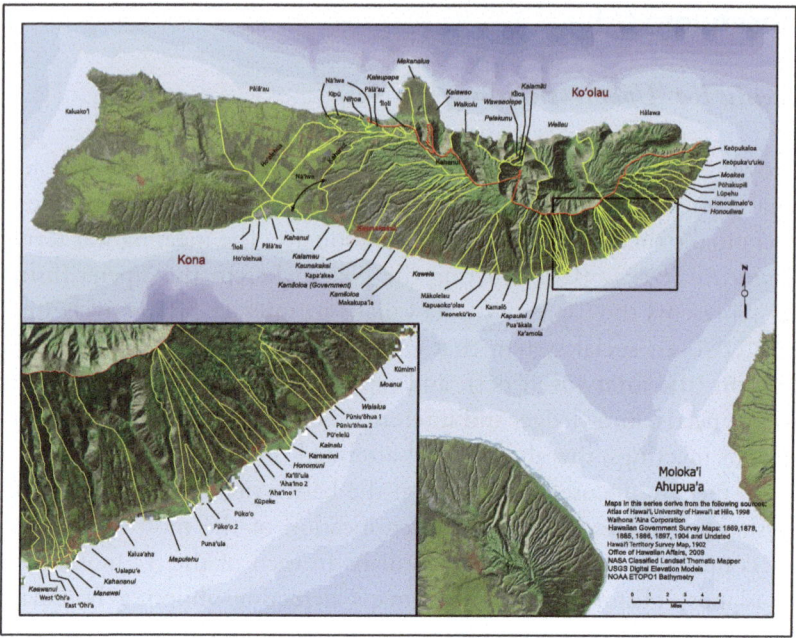

Fig. 4.5 Ahupuaʻa and moku of Molokaʻi

traditional boundaries is provided below (Fig. 4.5). Participants adopted the *Hoʻohanohano I Nā Kūpuna Puwalu* series mission statement to serve as inspiration for any future work associated with the ʻAha Moku process.

The relationship between the Hawaiian people and natural resources of land and sea was said, during the ʻaha, to be symbiotic; that is, the health of one is dependent on the other. As noted throughout this text, the evolving particulars of ecological knowledge have been handed down over generations. This is a profound fact, the importance of which is difficult for newly arriving residents to fully grasp. Mr. Timmy Paulokakeioku Bailey, respected mahiʻai from Maui, stated that, "as Native Hawaiians, it is not our right, but our duty, to continue what our ancestors have set forth for us to proudly claim—we are Hawaiians with inherent rights and duties and today it is imperative that we at once understand our past culture and recognize our living culture."

The first day of the ʻaha was dedicated to identifying traditional moku and ahupuaʻa boundaries, and to discussing traditions associated with use of natural resources in adjacent ocean, shoreline, and mountains. Lūnamaka'āinana o mokupuni (island representatives) gave in-depth presentations about various moku and associated traditions on each island, and each discussed prevalent concerns about the status of local and regional natural resources. Priority topics were brought to the forefront, including the following:

- Practices and protocols needed to represent local and regional interests and concerns through the proposed standing councils;
- Immediate problems associated with shoreline development, such as increasingly limited public access to the ocean;
- Ongoing concerns that inter-generational communication of traditional Hawaiian values could fail in coming years;
- Persistent problems associated with indigenous fishing opportunities and rights;
- Problematic interactions between endemic and invasive marine and terrestrial species; and
- General and specific concerns regarding the health of the ocean, watersheds, streams, and rivers across the islands (Fig. 4.6).

Meeting participants reiterated the perspective that the 'Aha Moku was the most appropriate venue through which local experts, educators, and cultural practitioners could positively influence natural resource policy decisions around the Hawaiian Islands.

During the second day of the meeting, participants delved deep into practical issues associated with establishing the 'Aha Moku system. Criteria and qualifications for prospective representatives were discussed at length. It was determined that criteria should maximize representation of Native Hawaiian values in general while also accomodating variability in local needs and interests. The capacity to facilitate effective communication among all parties was deemed critical. It was determined that 'Aha Moku members would need to meet criteria approved by island representatives present at the meeting, but that processes for selecting participants in a given 'Aha Ahupua'a or 'Aha Moku could and should vary by island as indicated by historic processes and/or the contemporary needs and interests of those being represented.

Fig. 4.6 'Auwai (irrigation systems) to water kalo (taro) patches in the ahupua'a of Mānoa on O'ahu are centuries old (Photo by Sylvia Spalding)

Certain meeting attendees offered their perspectives on the ideal nature and subjects of representation. It was agreed that the system for representing local concerns and interests must in all cases be holistic in nature, capable of addressing management and use of land, water, shoreline, and ocean resources. The importance of clean ea (air) and the value of apoālewa (the space above, where the birds soar) as essential natural resources to be cared for by people in any given ahupua'a were also discussed. In the end, there was strong agreement about the need for a formal, legally sanctioned organization of standing councils for each moku on each of the main islands, and meeting participants agreed that persons with recognized expertise in lawai'a and mahi'ai should contribute in a meaningful way. The most popular sentiment was that the moku-level councils would represent the perspectives of participants in existing or prospective 'Aha Ahupua'a. The meeting ended with strong group affirmation to further any and all efforts to advance the overall 'Aha Moku process.

E Hoʻoni I Na Kai ʻEwalu! E Hoʻale Ka Lepo Popolo! (Stir up the Eight Seas, Rise up People of Hawaiʻi)

A fifth meeting was convened during October 2007 to consider and address practical issues and challenges associated with the ʻAha Ahupuaʻa, ʻAha Moku, and ʻAha Kiole. The first day of the meeting involved a series of breakout workshop sessions focused on ways and means for integrating ʻAha Moku representation into existing governmental and community programs that address or advocate sustainable use of natural resources and ecosystems around the Hawaiian Islands.

A variety of basic principles and best practices models for managing natural resources were discussed, and there was a natural tendency for traditional indigenous knowledge and practices to be considered most intensively. It was agreed that whatever principles or models would be adapted should reflect the predominant experiences and perspectives of kamaʻāina residing in the places in question. For example, people in some districts are well-versed or primarily interested in the harvest and management utility of the lunar calendar, others in seasonal kapu, and yet others in the more specific teachings of local kūpuna. Meeting participants generally agreed that while Hawaiian lineage is an important part of ensuring representation of Native Hawaiian needs, interests, and values, it would be beneficial to consider the experiences and perspectives that persons of other ancestries might also bring to the process.

The topic of traditional knowledge was woven through discussions during both days of the meeting, with emphasis on the desirability of residents as a whole to retain knowledge and traditions that are unique to each district and each island. In terms of who would best represent local needs, interest, and knowledge, it was determined that participants from each moku would be responsible for identifying persons with valid knowledge of (a) salient local resource issues; (b) kapu or other kānāwai or forms of governance specific to certain areas; and (c) viable means for enforcing customs, rules, and regulations on a local and regional basis.

Other topics covered during the discussion centered on mechanisms for ensuring that the specific needs and concerns of residents in each ahupuaʻa be clearly known and communicated upward to the regional councils. It was agreed that this would require educating residents about the ʻAha Moku process and opportunities for local representation in that process. Participants agreed that a repository of contact information would allow participants at all council levels to easily reach and work with ʻōhana and cultural specialists across the many moku and ahupuaʻa.

There was consensus that various assemblies will succeed to the extent that wise and effective leaders are involved at each level of the process. Among other attributes, it was said that leaders and representatives must possess traditional knowledge that has been handed down over the generations along with knowledge of spiritual connections between natural resources and Native Hawaiian values and practices.

Prospective consultation processes were discussed in relation to ʻAha Ahupuaʻa, ʻAha Moku, and the ʻAha Kiole. Among other attributes, it was determined that the ʻAha Moku and ʻAha Kiole would be accountable to ʻAha Ahupuaʻa in all matters, and that the latter would be autonomous community organizations. Moreover, it was determined that those persons being represented in a given ahupuaʻa may choose to manage natural resources based on one or a combination of traditional means: the konohiki system, kūpuna knowledge, seasonal kapus or other culturally appropriate approaches. Finally, it was decided that government agencies must be held accountable for actions that have or will impact traditional cultural and natural resources, but that it would ultimately be the responsibility of people in each ahupuaʻa to mālama (care for) local resources on a routine basis.

Necessary and appropriate functions of the ʻAha Moku were also discussed, and it was decided that the body should function as: (a) the facilitator for interactions between ʻAha Ahupuaʻa and government agencies, and the point of liaison and contact between ʻAha Ahupuaʻa and the ʻAha Kiole; (b) an integral part of existing governmental assessment and permitting processes and sources of information about these processes as they might affect each ahupuaʻa and moku; and (c) a source of mediation for resolving issues of pertinence to the various ʻAha Ahupuaʻa.

Finally, workshop attendees worked to define the role of the ʻAha Kiole. It was decided that the ʻAha Kiole should function to: (a) facilitate ʻAha Ahupuaʻa interactions with county, state, federal, and international agencies and issues; (b) seek a permanent seat on all governing boards and commissions that render decisions that could affect life in the ahupuaʻa and moku; (c) work to implement statutes and ordinances deemed necessary by the ʻAha Moku; (d) facilitate training and education necessary to assess and monitor natural resources in each ahupuaʻa and/or moku; (e) facilitate the training and education necessary to enable local enforcement of natural and cultural resource management strategies in each ahupuaʻa and moku; and (f) seek the cooperation of county, state, and federal agencies to aid in the enforcement of natural and cultural resource management strategies implemented in each ahupuaʻa and moku.

Participants ultimately identified five key objectives that would need to be continually considered as the ʻAha process advances over time. These were revisited during subsequentmeetings and are summarized here:

- **Connect the Ocean and Land**: Deeply consider and value the connection of land and sea; the importance of healthy wetlands, streams, and estuaries and how these impact the health of island ecosystems; and maintain attention to factors and processes that impact the shoreline, marine habitats, and fisheries;
- **Safeguard Native Hawaiian Traditional Resource Methodology and Sustainability**: Strive to protect the cultural and natural resource traditions and customs of Native Hawaiians;
- **Continue to Streamline an Administrative Structure for the ʻAha Moku**: Develop media and public relations programs to educate all islanders about the merits of traditional means of stewardship;
- **Promote Collaboration, Education and Stewardship**: Identify specific resources and traditional methodologies employed in the sustainable use of natural resources of land and sea; establish a consensus process on natural resource use issues and management strategies based on Native Hawaiian knowledge and traditions; build capacity for community participation in traditional use and management of natural resources; establish means for effective interaction with educational facilities such as public, private, charter and vocational schools and universities; and establish an information repository; and
- **Institutionalize Integrated Natural and Cultural Resource Management**: Develop legislative and administrative proposals to

improve management of natural resources; establish seats on relevant government committees; and develop direct person-to-person links between representatives of ʻAha Moku and state and federal government agencies.

In keeping with these objectives, more than 80 community meetings were held across the islands between 2007 and 2009 to gather testimony from cultural practioners, kūpuna, and others with an interest in traditional use and management of natural resources. Coupled with the results of *Hoʻo Lei Ia Pae ʻĀina Puwalu* (described below), the input so received formed a vital part of a 2010 report developed by the ʻAha Kiole Advisory Committee that had been formed to advance the overall ʻAha Moku effort before its sunsetting in 2011 (see ʻAha Kiole Advisory Committee 2010).

Hoʻo Lei Pae ʻĀina Puwalu (Cast the Net, Bring All Together in Hawaiʻi)

A sixth ʻaha was held in November 2010 to further refine formal recommendations for enabling the input of the Native Hawaiian community into decisions regarding the use of the state'spublic trust natural resources. As discussed throughout the previous meetings, this is to be accomplished by reestablishing traditional place-based consultation processes and relating these to the existing system of governance. This goal was furthered during the sixth meeting, as participants crystallized recommendation regarding the structure and objectives of the ʻAha Kiole and ʻAha Moku, and as the group refined its understanding of and recommendations for best practices of natural resource management. The meeting involved review of modern resource management challenges by key persons in both the public and private sectors, and deliberation on how traditional knowledge, values, and practices could be used to prioritize and solve such challenges. Among the most significant discussions was that identifying the "five pillars" that underlay incorporation of a traditional ahupuaʻa-moku structure of resource management in the modern context:

1. **Adaptive Management**. The ancient system is adaptive in nature–
   problems are addressed through a negotiated district-wide rule-mak-
   ing process. The effects of rules, such as seasonal kapu on the take
   of certain species are observed and evaluated. If such measures are
   deemed effective, they are upheld over time. If they are not proven
   effective, they are modified appropriately.
2. **Code of Conduct**. Community-based social controls on conduct are formulated to support the adaptive management strategy.
   Representatives from each moku determine an appropriate code of
   conduct through which people can wisely use and conserve resources
   under their kuleana.
3. **Community Consultation**. Resource managers interact with community representatives who know the resources, people, issues, and
   ahupua'a in question. A respected individual facilitates communication between the community and the resource managers.
4. **Educational Programs**. Traditional use and management of natural
   resources, resource monitoring, and adaptive management strategies can be learned through the existing educational system, through
   inter-generational communication of knowledge, and through
   hands-on experience. Communities possess the capacity to teach
   managers about local resources and time-tested, place specific management strategies and community, group, and other collective needs
   and interests.
5. **Eligibility Criteria**. Knowledge is central to effective management
   of natural resources, and it is often the case that certain residents in
   a given ahupua'a are particularly knowledgeable about the natural
   environment. Traditional ecological knowledge is often of great utility for resource managers, and it can be developed in relation to a
   variety of social and ecological contexts (Fig. 4.7).

Finally, meeting participants refined and finalized a resolution intended
to further the values, needs, and critical interests of the indigenous people of Hawai'i through the 'Aha Moku process as it is undertaken in

Fig. 4.7 Wahi Kapu, Moloka'i

years to come. As provided below, this was developed for presentation and furtherance of the mission during the 2011 session of the Hawai'i State Legislature.

Efforts to advance the content of the resolution were fruitful, and on July 9, 2012, Act 212 as amended was signed into law. As stated in the Act itself (Fig. 4.8):

> Between 2006 and 2010, three more puwalu [or 'aha, beyond the *Ho'ohanohano I Nā Kūpuna Puwalu* series] were convened to gather additional community input on best practices in the area of native Hawaiian resource management. All puwalu were open to the public and included farmers, fishers, environmentalists, educators, organizations and agencies, and governmental representatives who, through discussions on the integration of these practices into regulation and common utilization, came to the consensus of the necessity of integrating the aha moku system into government policy. The information gathered from all puwalu has been compiled into annual comprehensive reports to the legislature as required by Act 212, Session Laws of 22 Hawaii 2007, as amended by Act 39, Session Laws of Hawaii 2009. The purpose of this Act is to formally

URGING THE COUNTY, STATE AND FEDERAL ENTITIES TO FORMALLY RECOGNIZE THE AHA MOKU SYSTEM AS PART OF THE MANAGEMENT REGIME OF NATURAL RESOURCES IN HAWAII

WHEREAS, the statewide Ho`o Lei `Ia Pae `Āina Puwalu was held at the Hawaii Convention Center in Honolulu on November 19 and 20, 2010, involving more than 200 native Hawaiian traditional practitioners, fishermen, farmers, ranchers, educators, municipal representatives, State representatives and the general public, and

WHEREAS, it was agreed that the `Aha Moku structure is a traditional, effective, and community-based way to manage natural resources in Hawai`i, and

WHEREAS the island caucuses at the Puwalu agreed that the Hawaii State Legislature should extend, amend and implement Act 212:

- That the `Aha Moku system be continued;
- That the recommendations from each island in the 2009 `Aha Kiole report to the Legislature be implemented;
- That new `Aha Kiole representatives be selected/elected by `Aha Moku councils that have been established on each of the mokupuni;
- That where Aha Moku councils have not yet been established, efforts be made to establish them as soon as possible;
- That Niihau a Kahelelani continue to be managed based on and exclusively under its konohiki system;
- That these Aha Moku councils be formally recognized;
- That the `Aha Kiole role be amended so as to include it being the conduit between the Aha Moku system and the State of Hawaii Legislature; and
- That the new `Aha Kiole report back to the Legislature on the status of the `Aha Moku system throughout the pae`aina at the end 2011, and

WHEREAS the Puwalu participants also supported customary and traditional practices that have sustained the native Hawaiian population and culture, such as the cultural, non-commercial take of *honu* and fish from waters throughout the Hawaii Archipelago;

NOW, THEREFORE, BE IT RESOLVED that the participants of the Hawai`i Statewide Puwalu, in conference at the Hawai`i Convention Center on November 19 and 20, 2010, urges the county, state and federal entities to formally recognize the `Aha Moku system as part of the management regime of natural resources in Hawaii and the allowance of customary and traditional practices; and

BE IT FURTHER RESOLVED that copies of this Resolution be transmitted in Hawaiian and English to all County Mayors, Governor of Hawaii, President of the Senate, Speaker of the House, Senate Committee on Judiciary and Hawaiian Affairs Chair, House Committee on Hawaiian Affairs Chair, Office of Hawaiian Affairs Board of Trustees Chair, Secretary of Commerce, the Chair of the Western Pacific Regional Fishery Management Council, and the US Ambassador to the United Nations.

recognize the aha moku system and to establish the aha moku advisory committee within the department of land and natural resources, which may serve in an advisory capacity to the chairperson of the board of land and natural resources. The aha moku advisory committee may advise on issues related to land and natural resource management through the aha moku system, a system of best practices that is based upon the indigenous resource management practices of moku (regional) boundaries, which

Fig. 4.8 Gov. Neil Abercrombie signs into law formal recognition of the 'Aha Moku system and establishment of the 'Aha Moku Advisory Committee, surrounded by members of the Native Hawaiian community who brought with them photos of 'Aha Moku supporters who had passed on (Photo courtesy of the Western Pacific Regional Fishery Management Council)

acknowledges the natural contours of land, the specific resources located within those areas, and the methodology necessary to sustain resources and the community. The aha moku system will foster understanding and practical use of knowledge, including native Hawaiian methodology and expertise, to assure responsible stewardship and awareness of the interconnections of the clouds, forests, valleys, land, streams, fishponds, and sea. The moku system will include the use of community expertise and establish programs and projects to improve communication, provide training on stewardship issues throughout the region (moku), and increase education. The establishment of this committee does not preclude any person's or organization's right to provide advice to the department of land and natural resources.

Thus, the 'Aha Moku Advisory Committee or AMAC was formally established in 2012 and emplaced in the State of Hawaii Department of Land and Natural Resources for purposes of administration, with the ongoing mission to advise the Chair of the Board of Land and Natural Resources on issues pertaining to use and management of natural and cultural resources around the islands. The numerous 'aha held between 2006 and 2011 were, therefore, successful in advancing the role of place-specific traditional knowledge and practical experience in the state's extant natural resource management process. Moreover, the overall 'Aha Moku process has generated a variety of locally positive effects irrespective of the formal interface with state government, not the least of which is heightened awareness of the present and future potential of traditional place-based resource management. As discussed in the concluding chapter of this text, the effort has thus far yielded a variety of concrete successes—but not without an extensive outlay of energy on the part of many highly committed individuals and groups, and not without significant challenges on the horizon.

References

'Aha Kiole Advisory Committee. (2010). *Best Practices and Structure for the Management of Natural and Cultural Resources in Hawaii*. Final Report. In response to Act 212 relating to Native Hawaiians, and Act 39 relating to Native Hawaiians. December. Honolulu. Available at http://ahamoku.org/wp-content/uploads/2011/09/2011-Aha-Kiole-Legislative-Report-Final.pdf.

Ahmed, M., Maclean, J., Gerpacio, R. V., & Sombilla, M.A. (2011). *Food Security and Climate Change in the Pacific—Rethinking the Options*. Pacific Studies Series. Mandaluyong City, Philippines: Asian Development Bank.

Baldauf, N. (2014). *Native Hawaiian Law—A Treatise, Chapter 16: Iwi Kūpuna—Native Hawaiian Burial Rights*. A collaborative effort of the Native Hawaiian Legal Corporation, Ka Huli Ao Center for Excellence in Native Hawaiian Law at the William S. Richardson School of Law—University of Hawai'i at Mānoa. Kamehameha Publishing, Honolulu.

Blendon, R. J., Miller, C., Gudenkauf, A., et al. (2017). *Discrimination in America: Experiences and View of Native Americans*. National Public Radio, Robert Woods Johnson Foundation, and Harvard's T. H. Chan School of Public Health. Available at https://www.npr.org/assets/img/2017/10/23/discriminationpoll-african-americans.pdf.

Brave Heart, M. (2003). The Historical Trauma Response Among Natives and Its Relationship with Substance Abuse: A Lakota Illustration. *Journal of Psychoactive Drugs, 35*(1), 7–13.

Brown-Rice, K. (2013). Examining the Theory of Historical Trauma Among Native Americans. *The Professional Counselor, 3*(3), 117–130.

Champagne, D. (2007). *Social Change and Cultural Continuity Among Native Nations.* Plymouth, UK: Altamira Press. A Division of Rowman and Littlefield Publishers, Inc.

Chong, C. (1996). Effect of Flow Regimes on Productivity in Hawaiian Stream Ecosystems. In *Will Stream Restoration Benefit Freshwater, Estuarine and Marine Fisheries? Proceedings of the Hawaii Stream Restoration Symposium* (pp. 152–157). October 1994. Honolulu: State of Hawaii, Division of Aquatic Resources and American Fisheries Society, Hawaii Chapter.

Chun, J., Chun, M.-C., & Liu, L. (2016). *2016 Annual Visitor Report.* Hawaii Tourism Authority Compiled by the Tourism Research Staff, G. Szigeti Director, Honolulu.

Chun Smith, A. A. (1994). *The Effects of Nutrient Loading on C:N:P Ratios of Marine Macroalgae in Kāneʻohe Bay, Hawaii.* Thesis completed in partial fulfillment of the requirements for the Master of Science degree, Department of Oceanography, University of Hawaii at Mānoa, Honolulu. 80 pp.

Churchill, W., & Venue, S. H. (2004). *Islands in Captivity—The International Tribunal on the Rights of Indigenous Hawaiians.* Cambridge, MA: Southend Press.

Commission on Water Resource Management. (1995). *In the Matter of Water Use Permit Applications, Petitions for Interim Instream Flow Standard Amendments, and Petitions for Water Reservations for the Waiāhole Ditch Combined Contested Case Hearing* (CCH-OA-95-1). Prepared by the Hawaii Commission on Water Resource Management. Department of Land and Natural Resources, State of Hawaii, Honolulu.

Cook, B. P., Withy, K., & Tarallo-Jensen, L. (2003). Cultural Trauma, Hawaiian Spirituality, and Contemporary Health Status. *Californian Journal of Health Promotion.* Special Issue: Hawaii, *1*, 10–24.

DaMate, L. (2016). Interview. ʻAha Moku- John Kaʻimikauaʻs vision. https://vimeo.com/157694988.

Danieli, Y. (1998). *International Handbook of Multigenerational Legacies of Trauma.* New York: Plenum Press.

Daniels, J. (2018 Winter). The Algorithmic Rise of the "Alt-Right." *Contexts.* Sage Journals Publication. https://doi.org/10.1177/1536504218766547. Available at http://journals.sagepub.com/doi/abs/10.1177/1536504218766547.

Dippie, B. (1982). *The Vanishing American: White Attitudes and U.S. Indian Policy*. Middleton, CT: Wesleyan University Press.

Duponte, K., Martin, T. Mokuau, N., & Paglinawan, L. (2010, February). 'Ike Hawai'i—A Training Program for Working with Native Hawaiians. *Journal of Indigenous Voices in Social Work, 1*(1), 1–24.

Duran, E., & Duran, B. (1995). *Native American Postcolonial Society*. Albany: State University of New York Press.

Englund, R. A., & Filbert, R. (1996, October). Relationship Between Stream Alteration and Fish Species Composition in O'ahu Streams. In *Will Stream Restoration Benefit Freshwater, Estuarine and Marine Fisheries? Proceedings of the Hawaii Stream Restoration Symposium* (p. 158). Honolulu: State of Hawaii, Division of Aquatic Resources and American Fisheries Society, Hawaii Chapter.

Englund, R. A., Arakaki, K., Preston, D. J., Coles, S. L., & Eldredge, L. G. (2000). *Nonindigenous Freshwater and Estuarine Species Introductions and Their Potential to Affect Sportfishing in the Lower Stream and Estuarine Regions of the South and West Shores of O'ahu, Hawaii* (Technical Report No. 17). Honolulu: The State of Hawaii, Department of Land and Natural Resources, Division of Aquatic Resources. Bernice Pauahi Bishop Museum. 121 pp.

Evans, C., Maragos, J., & Holthus, P. (1986). Reef Corals in Kāne'ohe Bay—Six Years Before and After Termination of Sewage Discharges. In P. L. Jokiel, R. H. Richmond, & R. A. Rodgers (Eds.), *Coral Reef Population Biology* (pp. 76–90) (Technical Report Number 37//U.S. Sea Grant Cooperative Report Number UNIHISEAGRANT-CR-86-01). Kaneohe: University of Hawaii Institute for Marine Biology, 501 pp.

Fabricius, K. E., De'ath, G., Noonan, S., & Uthicke, S. (2013). Ecological Effects of Ocean Acidification and Habitat Complexity on Reef-Associated Macroinvertebrate Communities. *Proceedings of the Royal Society—Biological Sciences, 281*(1775), 20132479. https://doi.org/10.1098/rspb.2013.2479.

Ferguson Wood, E. J., & Johannes, R. E. (Eds.). (1975). *Tropical Marine Pollution* (192 pp). Amsterdam and Oxford, NY: Elsevier Scientific Publishing Company.

Fogelman, C. (2018). *Demographic, Social, Economic, and Housing Characteristics for Selected Race Groups in Hawaii*. Honolulu: State of Hawaii, Research and Economic Analysis Division (READ) of the Department of Business, Economic Development & Tourism.

Font, W. F., Tate, D. C., & Llewellyn, D. W. (1996). Colonization of Native Hawaiian Stream Fishes by Helminth Parasites. In *Will Stream Restoration*

Benefit Freshwater, Estuarine and Marine Fisheries? Proceedings of the Hawaii Stream Restoration Symposium (pp. 94–111). October 1994. Honolulu: State of Hawaii, Division of Aquatic Resources and American Fisheries Society, Hawaii Chapter.

Friedlander, A., & Brown, E. (2004). Marine Protected Areas and Community-Based Fisheries Management in Hawaii. In *Status of Hawaii's Coastal Fisheries in the New Millenium. 2004 revised edition. Proceedings of the 2001 Fisheries Symposium sponsored by the American Fisheries Society*, Hawaii Chapter. A. M. Friedlander (ed.).

Friedlander, A. M., DeFelice, R. C., Parrish, J. D., & Frederick, J. L. (1997). *Habitat Resources and Recreational Fish Populations at Hanalei Bay, Kauai.* Final Report Submitted to the State of Hawaii, Department of Land and Natural Resources, Division of Aquatic Resources. U.S. Fish and Wildlife Service and the University of Hawaii, Hawaii Cooperative Fishery Research Unit, Honolulu, 320 pp.

Friedlander, A. M., Nowlis, J., & Koike, H. (2015). Improving Fisheries Assessments Using Historical Data. In J. N. Kittinger, L. McClenachan, K. Gedan, & L. K. Blight (Eds.), *Marine Historical Ecology—Applying the Past to Manage for the Future.* Oakland, CA: University of California Press.

Friedlander, A. M., Parrish, J. D., & DeFelice, R. C. (2002). Ecology of the Introduced Snapper *Lutjanus kasmira* (Försskal) in the Reef Fish Assemblage of a Hawaiian Bay. *Journal of Fisheries Biology, 60,* pp. 28–48.

Galinsky, A. M., Zelaya, C. E., Simile, C., & Barnes, P. M. (2014). *Health Conditions and Behaviors of Native Hawaiian and Pacific Islander Persons in the United States, 2014.* Hyattsville, MD: U.S. Department of Health and Human Services, Centers for Disease Control and Prevention, National Center for Health Statistics. Series 3, Number 40. DHHS Publication No. 2017-1424.

Glazier, E. W. (2007). *Hawaiian Fishermen.* Belmont, CA: Wadsworth-Cengage Publishers.

Glazier, E. W. (Ed.). (2011). *Ecosystem Based Fisheries Management in the Western Pacific.* Hoboken, NJ: Wiley-Blackwell Publishers. ISBN 978-0-8138-2154-2.

Glazier, E. W., Carothers, C., Milne, N., & Iwamoto, M. (2013). Seafood and Society on Oʻahu in the Main Hawaiian Islands. *Pacific Science, 67*(3). In Special Issue of *Pacific Science*—Human Dimensions of Small-Scale and Traditional Fisheries in the Asia-Pacific Region (J. Kittinger & E. W. Glazier, Eds.).

Goo, S. K. (2015). *After 200 Years, Native Hawaiians Make a Comeback.* Washington, DC: Pew Research Center. Available at http://www.pewresearch.org/fact-tank/2015/04/06/native-hawaiian-population/.

Grafeld, S., Oleson, K. L. L., Teneva, L., & Kittinger, J. N. (2017). Follow That Fish: Uncovering the Hidden Blue Economy in Coral Reef Fisheries. *PLoS ONE, 12*(8), e0182104. https://doi.org/10.1371/journal.pone.0182104.

Grigg, R. W. (1972). *Some Ecological Effects of Discharged Sugar Mill Wastes on Marine Life.* Honolulu: Water Resources Research Center, University of Hawaii at Manoa. Water Resources Seminar Series 2, pp. 27–45.

Grigg, R. W. (1994). Effects of Sewage Discharge, Fishing Pressure and Habitat Complexity on Coral Ecosystems and Reef Fishes in Hawaii. *Marine Ecology Progress Series, 103,* 25–34.

Grigg, R. W. (1995). Coral Reefs in an Urban Environment in Hawaii: A Complex Case History Controlled by Natural and Anthropogenic Stress. *Coral Reefs, 14,* 253–266.

Harman, R. F., & Katekaru, A. Z. (1988). *Hawaii Commercial Fishing Survey 1987* (71 pp.). Honolulu: State of Hawaii, Department of Land and Natural Resources, Division of Aquatic Resources.

Hostetler, C. J. (2014). Income Inequality and Native Hawaiian Communities in the Wake of the Great Recession: 2005–2013. Ho'okahua Waiwai, Economic Self-Sufficiency Fact Sheet, Volume 2014. Number 2. Office of Hawaiian Affairs, Research Division, Special Projects, Honolulu.

Hunter, C. L., & Evans, C. W. (1995). Coral Reefs in Kāne'ohe Bay, Hawaii: Two Centuries of Western Influence and Two Decades of Data. *Bulletin of Marine Science, 57,* 501–515.

Inafuku, J. K. (2015). E Kūkulu ke Ea: Hawai'i's Duty to Fund Kaho'olawe's Restoration Following the Navy's Incomplete Cleanup. *Asian-Pacific Law and Policy Journal, 16*(2), 22–69.

Itano, D. G., & Holland, K. N. (2000). Movement and Vulnerability of Bigeye (Thunnus obesus) and Yellowfin Tuna (Thunnus albacares) in Relation to FADs and Natural Aggregation Points. *Aquatic Living Resources, 13,* 213–223.

Johannes, R. E. (1982). Traditional Conservation Methods and Protected Marine Areas in Oceania. *Ambio, 11*(5), 258–261.

Johnson, A., & Vertessy, R. (2014). Preface to Climate Variability, Extremes and Change in the Western Tropical Pacific: New Science and Updated Country Reports. Prepared by the Australian Bureau of Meteorology and Commonwealth Scientific and Industrial Research Organisation (CSIRO), Pacific-Australia Climate Change Science and Adaptation Planning

Program Technical Report. Abbs, D., Grose, M., Hemer, M., Lenton, A., & McGee, S. (lead authors). Melbourne.

Jokiel, P. L., Rodgers, K. S., Walsh, W. J., Polhemus, D. A., & Wilhelm, T. A. (2011). Marine Resource Management in the Hawaiian Archipelago: The Traditional Hawaiian System in Relation to the Western Approach. *Journal of Marine Biology, 2011*, 1–16.

Kaneohe Bay Master Plan Task Force. (1992). *Kāneʻohe Bay Master Plan* (115 pp.). Prepared by the Kāneʻohe Bay Master Plan Task Force. Honolulu: State of Hawaii, Office of State Planning.

Kaʻaiʻai, C. (2016, October). Wetern Pacific Indigenous Fishing Communities. *Pacific Islands Fishery Monographs.* A Publication of the Western Pacific Regional Fishery Management Council. Number 7.

Kamehameha Schools. (2014). *Kalo kanu o ka ʻāina. In Ka Huakaʻi: 2014 Native Hawaiian Educational Assessment.* HonoluluKamehameha Publishing.

Kanaʻiaupuni, S. M., & Malone, N. (2006). This Land Is My Land: the Role of Place in Native Hawaiian Identity. *Hūlili: Multidisciplinary Research on Hawaiian Well- Being, 3*(1), pp. 281–307.

Kaufman, L. (1986). Why the Ark Is Sinking. In L. Kaufman & K. Mallory (Eds.), *The Last Extinction* (pp. 1–41). Cambridge and London: Massachusetts Institute of Technology Press, 208 pp.

Keala, G. G., Hollyer, J. R., & Castro, L. (2007). *Loko Iʻa—A Manual on Hawaiian Fishpond Restoration and Management.* College of Tropical Agriculture and Human Resources. Honolulu: University of Hawaiʻi at Mānoa.

Kittinger, J., Teneva, L. T., Koike, H., Stamoulis, K. A., Kittinger, D. S., Oleson, K. L., et al. (2015). From Reef to Table: Social and Ecological Factors Affecting Coral Reef Fisheries, Artisanal Seafood Supply Chains, and Seafood Security. *PLoS ONE, 10*(2015), e0123856. https://doi.org/10.1371/journal.pone.0123856.

Kittinger, J. N., Koehn, J. Z., Le Cornu, E., Ban, N. C., Gopnik, M., Armsby, M., et al. (2014). A Practical Approach for Putting People in Ecosystem-Based Ocean Planning. *Frontiers in Ecology and the Environment.* http://dx.doi.org/10.1890/130267.

Kosaki, R. H. (1954). *Konohiki Fishing Rights* (Report Number 1). Honolulu: Legislative Reference Bureau.

Laws, E. A., & Allen, C. B. (1996). Water Quality in a Subtropical Embayment More Than a Decade After Diversion of Sewage Discharges. *Pacific Science, 50,* 194–210.

Leary, J. D. (2005). *Post Traumatic Slave Syndrome: America's Legacy of Enduring Injury and Healing*. Portland, OR: Uptone Press.

Lechner, A., Cavanaugh, M., & Blyler, C. (2016). *Addressing Trauma in American Indian and Alaska Native Youth*. Final Report. Mathematica Policy Research Report. Prepared for the U.S. Department of Health and Human Services, Assistant Secretary for Planning and Evaluation. D. Chollet, Project Director. Reference Number 40146, Washington, DC.

Lindsey, R. H. K. (2002, Summer). Akaka Bill: Native Hawaiians, Larger Realities, and Politics as Usual. *University of Hawaii Law Review, 24*, 693–727.

Liu, D. M. K. (2005). E ʻao luʻau a kualima: Writing and Rewriting the Body and the Nation. *Californian Journal of Health Promotion, 3*(4), pp. 74–104. University of Hawaii Internal Medicine/Pediatrics Program.

Liu, D. M. K., & Alameda, C. K. (2011). Social Determinants of Health for Native Hawaiian Children and Adolescents. *Hawaiʻi Medical Journal, 70*, 9–14.

Loppie, S., Reading, C., & de Leeuw, S. (2014). Aboriginal Experiences with Racism and Its Impacts. *Social Determinants of Health*. National Collaborating Centre for Aboriginal Health. University of British Columbia. Prince George. Available at https://www.ccnsa-nccah. ca/docs/determinants/FS-AboriginalExperiencesRacismImpa cts-Loppie-Reading-deLeeuw-EN.

Lowe, K. (1996). Concerns for Coastal and Inshore Fisheries of Kāneʻohe Bay and the Nearby Windward Coast If Stream Flows Are Restored in Northern Kāneʻohe Bay. In *Will Stream Restoration Benefit Freshwater, Estuarine and Marine Fisheries? Proceedings of the Hawaii Stream Restoration Symposium* (pp. 31–56). October 1994. Honolulu: State of Hawaii, Division of Aquatic Resources and American Fisheries Society, Hawaii Chapter.

Lowe, K. (2004). The Status of Inshore Fisheries Ecosystems in the Main Hawaiian Islands at the Dawn of the Millenium: Cultural Impacts, Fisheries Trends and Management Challenges. In *Status of Hawaii's Coastal Fisheries in the New Millennium*. Revised edition. A. M. Friedlander (ed.). *Proceedings of the 2001 Fisheries Symposium*. Sponsored by, The American Fisheries Society and Hawaiʻi Chapter and Hawaiʻi Community Foundation. Prepared by the Hawaiʻi Audubon Society Hawaiʻi Department of Land and Natural Resources, Division of Aquatic Resources Hawaiʻi Cooperative Fishery Research Unit, University of Hawaiʻi, Honolulu.

Maly, K., & Maly, O. (2003). *Ka Hana Lawaiʻa A Me Na Koʻa O Na Kai ʻEwalu: A History of Fishing Practices and Marine Fisheries of the Hawaiian Islands* (Vols. I and II). Hilo: Kumu Pono Associates. Prepared for the Nature Conservancy and Kamehameha Schools.

Maragos, J. E., & Grober-Dunsmore, R. (Eds.). (1999). Proceedings of the Hawaii Coral Reef Monitoring Workshop. June 9–11, 1998. University of Hawaii East-West Center and State of Hawaii, Department of Land and Natural Resources (Publishers). Honolulu, Hawaii, 334 pp.

Matsuoka, J., McGregor, D., Minerbi, L., & Akutagawa M. (Eds.). (1994). *Governor's Molokaʻi Subsistence Task Force Final Report.* State of Hawaiʻi. Honolulu: The Task Force.

McCoy K. S., Williams, I. D., Friedlander, A. M., Ma, H., Teneva, L., & Kittinger, J. N. (2018). Estimating Nearshore Coral Reef-Associated Fisheries Production from the Main Hawaiian Islands. *PLoS ONE, 13*(4), e0195840. Available at https://doi.org/10.1371/%20journal.pone.0195840.

McDougall, B. Nālani. (2010). From Ue to Kue: Loss and Resistance in Haunani-Kay Trask's Night Is a Shark Skin Drum and Matthew Kaopio's Written in the Sky. *Anglistica, 14*(2), 51–62. Special Issue: Sustaining Hawaiian Sovereignty (D. Izzo & B. Kamaoli Kuwada, Eds.).

McGregor, D. P. (2007). *Na Kua ʻAina: Living Hawaiian Culture.* Honolulu: University of Hawaii Press.

McMullin, J. (2005). The Call to Life: Revitalizing a Healthy Hawaiian Identity. *Social Science and Medicine, 61,* 809–820.

Meller, N. (1985). *Indigenous Ocean Rights in Hawaii.* Honolulu: University of Hawaiʻi Sea Grant College Program, Sea Grant Marine Policy and Law Report. University of Hawaiʻi at Mānoa, Honolulu.

Melrose, J., Perroy, R., & Cares, S. (2016). *Statewide Agricultural Land Use Baseline 2015.* Prepared for the Hawaiʻi Department of Agriculture by the University of Hawaiʻi at Hilo, Spatial Data Analysis & Visualization Research Lab. Hilo.

Mokuau, N. (2002). Culturally Based Interventions Among Native Hawaiians. *Public Health Reports, 117*(Supp. 1).

Mokuau, N. (2011). Culturally Based Solutions to Preserve the Health of Native Hawaiians. *Journal of Ethnic & Cultural Diversity in Social Work, 20,* 98–113.

Mokuau, N., DeLeon, P. H., Kaholokua, J. K., Soares, S., Tsark, J. U., & Haia, C. (2016, October 31). *Challenges and Promise of Health Equity for Native Hawaiians. Perspectives/Expert Voices in Health and Health Care. National Academy of Medicine* (Discussion Paper). Washington, DC.

Available at https://nam.edu/wp-content/uploads/2016/10/%20Challenges-and-Promise-of-Health-Equity-for-Native-Hawaiians.pdf.

Nakasone, R. (1995, Summer). *A Study of Leaf Litter Fauna of the Waikane-Waiāhole Watershed: A Point-in-Time Survey of Stream Microbiota*. National Science Foundation, REU Program. SOEST, Honolulu: University of Hawaii Mānoa, 24 pp.

NOAA Coral Reef Conservation Program. (2018). *Coral Reef Conditions—A Status Report for the the Hawaiian Archipelago*. Prepared by the NOAA Coral Reef Conservation Program (CRCP) and the University of Maryland Center for Environmental Science. Honolulu: NOAA Fisheries Pacific Islands Fisheries Science Center.

Novak, V. (1993). Hawaii's Dirty Secret—The Continuing Denial of Native Hawaiian Land Rights. In W. Churchill & S. H. Venue (Eds.), *Islands in Captivity—The International Tribunal on the Rights of Indigenous Hawaiians* (pp. 191–203). Cambridge, MA: Southend Press.

Oneha, M. F. M. (2001). *Ka mauli o ka ʻāina a he mauli kānaka*: An ethnographic study from an Hawaiian sense of place. *Pacific Health Dialog, 8,* 299–311.

Osorio, J. K. (2014). Hawaiian Souls: The Movement to Stop the U.S. Military Bombing of Kahoʻolawe. In N. Goodyear-Kaopua, I. Hussey, & E. K. Wright (Eds.), *Nation Rising: Hawaiian Movements for Life, Land, and Sovereignty*. Durham: Duke University Press.

Perez. R. (2016). Hawaiians at Risk: Healing Efforts Return to Roots. *Honolulu Star Advertiser*. January 11. Honolulu. Available at http://www.staradvertiser.com/2016/01/11/hawaii-news/hawaiians-at-risk-healing-efforts-return-to-roots/.

Pokhrel, P., & Herzog, T. A. (2014). Historical Trauma and Substance Use Among Native Hawaiian College Students. *American Journal of Health Behavior, 38*(3), 420–429.

Pooley, S. G. (1987). Recommendations for a Five-Year Scientific Investigation of the Marine Resources and Environment of the Main Hawaiian Islands. U.S. National Marine Fisheries Service, Southwest Fisheries Center, Honolulu Laboratory (US NMFS-SWFC/HL), Admin. Rep. H-98-02: 22 pp.

Pukui, M. K., & Elbert, S. H. (1986). *Hawaiian Dictionary* (6th ed.). Honolulu: University of Hawaii Press.

Ramirez, K. R. (1998). Healing Through Grief: Urban Indians Reimagining Culture and Community in San Jose, California. *American Indian Culture and Research Journal, 22*(4), 305–333.

Rodgers, Kuʻulei S., Jokiel, P., Brown, E. K., Hau, S., & Sparks, R. (2015). Over a Decade of Change in Spatial and Temporal Dynamics of Hawaiian Coral Reef Communities. *Pacific Science, 69*(1), 1–13.

Rooker, J. R., Wells, R. J. D., Itano, D. G., Thorrold, S. R., & Lee, J. M. (2016). Natal Origin and Population Connectiviety of Bigeye and Yellowfin Tuna in the Pacific Ocean. *Fisheries Oceanography, 25*(3), 277–291.

Sai, D. K. (2011). *Ua Mau Ke Ea—Sovereignty Endures: An Overview of the Political and legal History of the Hawaiian Islands.* Honolulu: Pūʻā Foundation.

Shomura, R. (1987). Hawaii's Marine Fisheries Resources: Yesterday (1900) and Today (1986). U.S. NationalMarine Fisheries Service, Southwest Fisheries Center, Honolulu Laboratory. NMFS-SWFC/HL. Admin. Rep. H-87-21: 14 pp.

Silbiger, N. J., Nelson, C. E., Remple, K., Sevilla, J. K., Quinlan, Z. A., Putnam, H. M. et al. (2018). Nutrient Pollution Disrupts Key Ecosystem Functions on Coral Reefs. In *Proceedings of the Royal Society, Biological Sciences.* Published 6 June. DOI: https://doi.org/10.1098/rspb.2017.2718. Available at http://rspb.royalsocietypublishing.org/content/285/1880/20172718.

Smith, S. V., Kimmerer, W. J., Laws, E. A., Brock, R. E., & Walsh, T. W. (1981). Kāneʻohe Sewage Diversion Experiment: Perspectives on Ecosystem Responses to Nutrient Perturbations. *Pacific Science, 35*, 270–395.

Sotero, M. M. (2006). A Conceptual Modal of Historical Trauma: Implications for Public Health Practice and Research. *Journal of Health Disparities Research and Practice, 1*, 93–108.

State of Hawaiʻi. (2010). *The Disparate Treatment of Native Hawaiians in the Criminal Justice System.* State of Hawaii, Office of Hawaiian Affairs. Prepared in cooperation with the Justice Policy Institute; the Myron B. Thompson School of Social Work at the University of Hawaii at Mānoa; Georgetown University; and the University of Hawaii at Mānoa Department of Urban and Regional Planning. Honolulu. Available at https://19of32x2yl33s8o4xza0gf14-wpengine.netdna-ssl.com/wp-content/uploads/2015/01/native-hawaiians-criminal-justice-system.pdf.

State of Hawaiʻi. (2018). *Haumea—Transforming the Health of Native Hawaiian Women and Empowering Wāhine Well-Being.* Honolulu: Office of Hawaiian Affairs.

State of Hawaii, Office of Planning. (2013). Hawaiʻi Ocean Resources Management Plan. Hawaii CZM Program. Coastal Zone Management.

A publication of the Hawaii Office of Planning, Coastal Zone Management Program, pursuant to National Oceanic and Atmospheric Administration Award Nos. NA09NOS4190120, NA11NOS4190095 and NA12NOS4190097, funded in part by the Coastal Zone Management Act of 1972, as amended, administered by the Office of Ocean and Coastal Resource Management, National Ocean Service, National Oceanic and Atmospheric Administration, United States Department of Commerce, Honolulu.

Teneva, L. T., Schemmel, E. & Kittinger, J. N. (2018). State of the Plate: Assessing Present and Future Contribution of Fisheries and Aquaculture to Hawai'i's Food Security. *Marine Policy*. Accepted April 18. https://doi.org/10.1016/j.marpol.2018.04.025.

Titcomb, M., & Pukui, M. K. (1951). Native Use of Fish in Hawai'i. Memoir 29. *Supplement to the Journal of the Polynesian Society* (Installment No. 1), 1–96.

Uchida, R. N., & Uchiyama, J. H. (Eds.). (1986). *Fishery Atlas of the Northwestern Hawaiian Islands*. U.S. Dept. of Commerce, National Oceanographic and Atmospheric Administration (NOAA), National Marine Fisheries Service (NMFS). NOAA Technical Report NMFS 38: 142 pp, Honolulu.

Underwood, J. G., Silbernagle, M., Nishimoto, M., & Uyehara, K. (2013). Managing Conservation Reliant Species: Hawai'i's Endangered Endemic Waterbirds. *PLoS ONE, 8*(6), e67872. Available at https://doi.org/10.1371/journal.pone.0067872.

U.S. Census Bureau. (2015). *2015 American Community Survey 5-Year Estimates*. Washington, DC.

U.S. Census Bureau. (2017). *Quick Facts: Honolulu County*. Washington, DC. Available at https://www.census.gov/quickfacts/fact/table/honolulucountyhawaii%23viewtop.

U.S. Census Bureau. (2018). *Quick Facts: Hawaii. Population Estimates Division*. Washington, DC: U.S. Department of Commerce. Available at https://www.census.gov/quickfacts/hi.

U.S. Commission on Human Rights. (2018, December). Broken Promises: Continuing Federal Funding Shortfall for Native Americans. Briefing Before the United States Commission on Civil Rights Held in Washington, DC. Available at https://www.usccr.gov/pubs/2018/12-20-Broken-Promises.pdf.

U.S. Department of Commerce, NOAA Fisheries. (2007). Magnuson-Stevens Fishery Conservation and Management Act. As Amended through January 12, 2007. As amended by the Magnuson-Stevens Fishery Conservation and Management Reauthorization Act (P.L. 109-479). An Act to provide for the conservation and management of the fisheries, and for other purposes. May. Second Printing. Washington, DC.

U. S. Geological Survey. (1999). Hawaii. U. S. Department of the Interior, U. S. Geological Survey. USGS Fact Sheet 012-99. May, 4 pp, Honolulu.

Vaughan, M. B. (2018). *Kaiāulu—Gathering Tides*. Corvallis: University of Oregon Press.

Vaughan, M. B., & Vitousek, P. M. (2013). Mahele: Sustaining Communities Through Small-Scale Inshore Fishery Catch and Sharing Networks. *Pacific Science, 67*(3), 329–344. In Special Issue of *Pacific Science*—Human Dimensions of Small-Scale and Traditional Fisheries in the Asia-Pacific Region (J. Kittinger & E. W. Glazier, Eds.).

Wedding L. M., Lecky, J., Gove, J. M., Walecka, H. R., Donovan, M. K., Williams, G. J. et al. (2018). Advancing the Integration of Spatial Data to Map Human and Natural Drivers on Coral Reefs. *PLoS ONE, 13*(3), e0189792. March. Available at https://doi.org/10.1371/journal.pone.0189792.

Wells, R. J. D., Rooker, J. R., & Itano, D. G. (2012). Nursery Origin of Yellowfin Tuna in the Hawaiian Islands. *Marine Ecology Progress Series, 461,* 187–196. https://doi.org/10.3354/meps09833.

Williams, I., Baum, J. K., Heenan, A., Nadon, M., & Brainard, R. (2015, April). Human, Oceanographic, and Habitat Drivers of Central and Western Pacific Coral Reef Assemblages. *PLoS ONE.* Available at https://doi.org/10.1371/journal.pone.0120516.

Yamamoto, M. N., & Tagawa, A. W. (2000). *Hawaiʻi's Native and Exotic Freshwater Animals* (p. 200). Honolulu: Mutual Publishing.

5

Concluding Discussion

5.1 Advancing the 'Aha Moku System

More than a thousand years after colonizing one of the most remote archipelagos on earth, and after centuries of resisting a continual procession of external agents and forces of detrimental change, indigenous Hawaiians in many walks of life continue to avidly pursue, use, and manage natural resources in extended family and community settings across the island chain. Traditional ecological knowledge continues to accrue and evolve in relation to ever-changing social and environmental factors and conditions. The resources themselves continue to provide: important sources of nutrition; means of cultural and spiritual expression, relaxation and recreation; and points of focus for the organization and functioning of indigenous island society and economy. In short, the use and management of natural resources and related sociocultural dimensions of Native Hawaiian society draw on a long Polynesian history, with great present-day cultural significance.

But as discussed in the previous chapter, problems and challenges of many sorts abound in the urban and rural settings of present-day Hawai'i, and constrain indigenous use and management of natural

© The Author(s) 2019
E. W. Glazier, *Tradition-Based Natural Resource Management*,
Palgrave Studies in Natural Resource Management,
https://doi.org/10.1007/978-3-030-14842-3_5

resources. In contradiction to inherent rights and privileges that would otherwise enable descendants of the original colonists to manage and use ancestral lands and adjacent fishing grounds as they deem appropriate, the course of history and modern economy, environment, and system of governance have limited indigenous agency to do so. In these respects, the *Ho'ohanohano I Nā Kupuna Puwalu* series and subsequent meetings described above were timely and much needed in their elicitation of broadly representative perspectives on many issues that are at the heart of Hawaiian culture. Like other conventions of Native Hawaiians held in days past, the meetings brought challenging issues to the fore and generated the collective momentum needed to advance tradition into the realm of formal legislation.

The meetings were unique both in terms of scope, convening representatives from moku (traditional districts) around the islands, and in terms of holistic treatment of interrelated topics and issues—from traditional ways of using and managing island ecosystems, to factors that impinge on such activities, to strategies for bringing traditional resource management strategies to the present. Of particular significance is the manner in which the meetings and subsequent legislation did not avoid particularly challenging issues but rather enabled a traditional approach for addressing them; accommodated variability in the way kama'āina (vernacular: one who has lived in Hawai'i for a long time) from disparate island locations understand, use, and manage living natural resources; and developed means for mediating conflicts that inevitably arise in the use and management of natural resources. These attributes were clearly stated in six resolutions generated during the well-attended puwalu called *Kānāwai Kai* (Proper Ways to Behave on the Ocean), convened in Honolulu in 2011 just prior to the passage of Act 288 by the Hawai'i State Legislature. The resolutions urged:

(1) Implementation of the 'Aha Moku system to resolve inter-island conflict; any protected species decision-making, rulemaking, conservation, recovery and management; any coastal and marine spatial planning in Hawai'i; and any review of fishery management processes in the State of Hawai'i;

(2) Resolution of inter-island conflicts through the 'Aha Kiole ['Aha Moku] system, which incorporates a lasting, fair, transparent and equitable process that respects the rights and needs of all parties in conflict;

(3) The federal government to ensure early and continued community engagement and consultation [as needed to enact equitable natural resource management actions and related assessment processes], which includes fully informed advice and consent by communities, as well as transparency, inclusion of the human and cultural environments, and timely action in any protected species decision-making, rulemaking, conservation, recovery, or management process;

(4) Implementation of the 'Aha Moku system as established under Act 212 and a formal consultation process with communities and agencies involved to ensure a fair and comprehensive review of the implementation and analysis of changes to any proposed National Marine Sanctuaries Act potentially affecting people in the State of Hawai'i;

(5) That the 'Aha Moku system should be the basis for conducting coastal and marine spatial planning in Hawai'i;

(6) A thorough review of fishery management in the State of Hawai'i with significant community involvement in the process to minimize duplication and [the human effects of] overly redundant regulations, and to enhance public compliance.

Three additional meetings have thus far been convened subsequent to the passage of Act 288 (State of Hawai'i 2012), each of which was intended to encourage the forward motion and efficiency of the 'Aha Moku system over time. *Lawelawe Hana Ke 'Aha Moku* (Serving the 'Aha Moku System) was held in West Maui during September 2014 to examine the nature of a working island council (that is, 'Aha Moku O Maui, the Maui council) and to encourage representatives of other 'Aha Moku island councils around Hawai'i to identify priority natural resource issues to be addressed, and to develop budgets for planned activities. Three resolutions were generated during this meeting:

(1) Ensure the inclusion of ʻAha Moku representatives on state and county commissions, boards, and advisory bodies [i.e., those mandated to address natural resource management issues in the islands];

(2) Urge the United States Congress to conduct an audit of the $400 million in funding provided under Title X of the 1994 Defense Appropriations Act for the clean-up and restoration of Kahoʻolawe Island; and

(3) Urge the Hawaiʻi State Legislature to amend Act 288 by implementing term limits for members of the ʻAha Moku Advisory Committee (AMAC), and establish minimum requirements for the Executive Director that include an understanding of state boards and commission standards, civil service requirements, and the operation and relationship of state agencies with regard to responsibilities of the AMAC to serve the people of Hawaiʻi.

Subsequently, on October 20, 2016, the AMAC adopted rules for practices and procedures to guide its work. The rules were adopted pursuant to SCR55-2015 (see State of Hawaiʻi 2016).

Ka Holomua ʻAna O Ka ʻAha Moku (Moving the ʻAha Moku System Forward) was held during November 2016 with an underlying focus on steps needed to ensure the forward momentum of the ʻAha Moku system. The meeting also functioned to (a) develop final procedural rules for the AMAC, (b) examine linkages and strategies for achieving mutually beneficial relationships between the respective island councils, and (c) review possibilities for involving youth in the ʻAha Moku process so as to ensure its perpetuation in the years to come. Two resolutions were generated during this meeting:

(1) Support the creation of opportunities for direct funding of the ʻAha Moku system at the ahupuaʻa, moku, and island levels [as available in state agency budgets, including those of] the Hawaii Tourism Authority and the Office of Hawaiian Affairs, in order to strengthen the organization and functionality of the ʻAha Moku system, including education of the kānaka (people), including ʻōpio (youth), in the knowledge and practices needed to sustain the system; and

(2) Recommend that the Hawaiian traditional practitioner seat on the [State of Hawai'i] Board of Land and Natural Resources [consistently] be filled by a Native Hawaiian.

Finally, a puwalu titled *Ola Honua I ke Kūpa'a Kanaka* (The Earth Flourishes with Bounty When the People Stand Together in Support) was convened during November 17–18, 2017. Meeting attendees sought to address prospective and actual challenges faced by the island councils, including organizational difficulties, funding needs, communication issues with the AMAC, and flagging momentum in certain island settings. Again, there was much emphasis on the need to involve youth in the 'Aha Moku process so that the hard-won ability to advise natural resource management decisions in the Hawaiian Islands would not be lost over time. This concern was stated by Walter Meheula Heen during the third conference in the *Ho'ohanohano I Nā Kupuna Puwalu*. Mr. Heen, whose illustrious career includes a variety of judicial appointments, tenure as State Senator, and Trustee for the Office of Hawaiian Affairs, among other accomplishments, spoke of the dire present and future need for Native Hawaiians to consistently "have a seat at the table—every table that affects Hawaiian culture, heritage, customs, and traditions" (Heen 2008: 20) (Figs. 5.1 and 5.2).

Fig. 5.1 Participants of the Ola Honua I Ke Kupa 'a Kanaka Puwalu (Photo courtesy of the Western Pacific Regional Fishery Management Council)

Fig. 5.2 Conference poster (Courtesy of the Western Pacific Regional Fishery Management Council)

5.2   Status to Date

Variability and Coalescence

Native Hawaiians share a common Polynesian ancestry and identity, a deep history as first-colonists of the remote island chain, a history of subtle and overt resistance to those bringing new problems to the islands, and the kūmulau (roots with many sprouts) of a rich island culture. But as for any complex society, there is much variation in perspective and life experience between individuals and families residing in any given island community or region. Moreover, because environmental conditions vary across the islands—perhaps especially between windward and leeward sides of any given island—there are notable differences in terms of: (a) how indigenous residents interact with the natural world and in terms of which resources are used and prioritized, (b) the knowledge that is developed through such interaction, and (c) perspectives on the best ways to address the challenges of managing natural resources in any given area.

Extant variability—in life experience and perspective; type and extent of traditional knowledge and practice; the nature of the local or regional environment; and many other factors, inevitably require that considerable deliberation be undertaken before group consensus can be attained on a given matter. Observation of the numerous puwalu described above made clear not only the nature and extent of such variability among indigenous attendees, but also the wealth of place-specific traditional knowledge and experience, and the impassioned nature of perspectives regarding what is needed to attain effective resource management in specific ahupua'a and moku around the islands.

Given many inter-island and within-island differences in context, and strong feelings about one's home ahupua'a and/or moku of residence (and/or those of one's larger family), it is to be expected that certain issues would take some time to resolve. For example, some indigenous groups on certain islands and in certain island districts were organized to address natural resources issues prior to the series of puwalu described in this text, while others coalesced, or are in the process of coalescing, in order to do so. As such, the ways in which 'Aha Moku system would ideally be organized and the rules and structure under which these would operate, ultimately required lengthy discussions over the course of multiple meetings. In the end, members of councils developing on the respective islands agreed to organize themselves in ways that would maximize the value of any preexisting indigenous natural resource-oriented organizations, human resources, accomplishments, and capabilities.

The Molokaʻi Example

The ʻAha Moku process is working well on the small, predominately indigenous-populated island of Molokaʻi (total pop. ~7345 in 2010, with ~4527 persons of Native Hawaiian ancestry in 2014; see Liliʻuokalani Trust 2018). The process has proceeded efficiently in part because of the extensive struggles, experience, and examples set by residents who have continually and proactively worked to protect natural resources and the people who depend on them (e.g., see profile of Malia Akutagawa at William S. Richardson School of Law 2018). A pre-existing state of readiness to undertake the basic tenets of the ʻAha Moku mission is also exemplified by the ongoing work of Hui Mālama O Moʻomomi, a perennially well-organized and active group that maintains kuleana (responsibility) for natural resource use and management in the Moʻomomi region of the rugged northwest Molokaʻi coastline.

Moʻomomi is widely considered a place of refuge for threatened endemic plant, avian, and other species, and it is also an important area for indigenous fishing and shoreline gathering activities. In conjunction with The Nature Conservancy, the State Department of Land and Natural Resources Natural Area Partnership Program, the Department of Hawaiian Homelands, and other agencies and entities, the local hui (organization) has advanced formal plans for a community-based subsistence fishing area in this remote island region, which has been used and managed by indigenous residents for many centuries (Hui Mālama O Moʻomomi 2017; see also Poepoe et al. 2003, 2007; Matsuoka et al. 1994) (Fig. 5.3).

In recent decades, not all island residents have been in full agreement with the traditional kānāwai (rules) employed for sustainable use and management of Moʻomomiʻs ocean and shoreline resources. These are based on the Hawaiian lunar calendar and evolving traditional knowledge of the area. But hui representatives and others have nevertheless exemplified what is required to protect and advance traditional natural resource-dependent ways of life in late twentieth and early twenty-first century Hawaiʻi. These requirements include:

Fig. 5.3 Mac Poepoe instructs students about traditional management of marine resources at Moʻomomi (Photo by Mark Mitsuyasu, Western Pacific Regional Fishery Management Council)

- Consistent observation and intergenerational transmission of knowledge regarding local ocean and island ecosystems and the natural resources nurtured by such ecosystems;
- Persistent application of local-traditional knowledge in regional fishing, hunting, farming, gathering, and aquaculture practices;
- Persistence in traditional use and stewardship of local ocean and island ecosystems and natural resources;
- A stern approach to dealing with people, factors, and processes that could detrimentally impact the ʻāina and thereby diminish traditional resource use and management practices (the majority of Molokaʻi residents have long been avidly opposed to land development projects that would not clearly benefit local society);
- Strong resistance to unwanted social and economic change; and

- Accrual of wisdom in the best approaches for enacting safeguards against misuse of land and sea, including development of outside partnerships with supportive governmental and non-governmental organizations.

Representative from the various moku councils on Moloka'i have managed to apply the 'Aha Moku process to pressing issues in real time, with the intent of advancing the overall well-being of island communities and residents. A case-in-point is the issue of cruise ship visitation at the small commercial harbor at Kaunakakai, the only town on the island (of a size that does not require a stoplight). The moku councils became involved after local protestors twice blocked access to moorings at Kaunakakai when a cruise ship arrived at the harbor unannounced. While many island residents feared the exigencies and potential environmental effects that large numbers of tourists arriving en masse could bring, local commercial vendors favored the prospective increase in business activity. In 2012, 'Aha Moku representatives seeking to examine and mediate the situation facilitated an island-wide survey to help gauge public perspectives on the cruise ship issue, while also querying persons attending regional 'Aha Moku meetings on the same matter. Some 85% of the 395 residents participating in the islandwide survey initially rejected cruise ship stopovers in their entirety. Some 11% welcomed the stopovers, and 4% stated these were acceptable only with various controls. Meanwhile, 36% of the 326 persons attending the 'Aha Moku meetings initially rejected the idea of Moloka'i stopovers entirely, 56% supported the prospective stopovers with adequate controls, and 8% supported uncontrolled stopovers (Cluett 2012).

In-depth negotiations between 'Aha Moku representatives, other local representatives, and the cruise ship firm ultimately resulted in a tightly controlled opportunity for the cruise ship company to moor a 36-passenger vessel at Kaunakakai 52 times each year (Hawaii News Now 2013). Agreed upon controls maximize the local business potential of the stopovers and minimize a wide range of potentially adverse effects on local society and environment. 'Aha Kiole O Moloka'i and the respective 'Aha Moku on this island continue to be active at the time of this writing.

On Other Islands

Similar to Moloka'i, Maui-based participants in the 'Aha Moku process have been particularly active to date, with productive relationships having been established with county, state, and federal government representatives to help examine and address a variety of pressing natural resource issues. Significantly, Maui participants have developed and incorporated an overarching standing council called 'Aha Moku o Maui, Inc., unique among the organizations established on other islands. Numerous 'Aha Moku initiatives continue to move forward on Maui, as guided by input from standing councils representing the moku of Lahaina, Wailuku, Hamakualoa, Hana, Kaupo, and Honua'ula working together under Aha Moku o Maui. Notably, a variety of natural resource committees have now been organized to address present and future challenges on the island as these relate to the natural and cultural resources of land, ocean, shoreline, air, water, and sacred burial sites.

A particularly useful product generated by a central figure in the 'Aha Moku process on Maui is the bilingual Hawaiian-English publication *Nānā I Ke Kumu* (Look to the Source). Created under the guidance of indigenous rights activist and federal employee Mr. Tim Bailey (Fig. 5.4), the publication describes the cultural and spiritual significance of what is now called Haleakalā National Park. The first of its kind in the 401-unit federal park system, the brochure, which also recommends steps needed to mālama (care for) this important upland region of Maui, was developed through extensive collaboration between local kūpuna, the National Park Service (NPS), and residents of the various ahupua'a encompassed by the park. Based on past rates of visitation, park officials anticipate that the brochure will be distributed to over one million persons each year. Funding for the project was provided by the non-profit Hawai'i Pacific Parks Association.

Progress has been made to advance the 'Aha Moku process on the Big Island, notable for its multitude of ahupua'a and extensive environmental variability between wet, lush windward coastlines, particularly dry leeward sides, and high mountain ecosystems (the summit of Mauna Kea is 13,803 feet above sea level). But environmental, sociopolitical, and other challenges, including the exigencies of travel between distant island locations, have thus far constrained full island-wide agreement and coalescence regarding the most effective manner of representation on the AMAC.

Similarly, the ʻAha Moku process is said to have progressed in fits and starts on both Oʻahu and Kauaʻi. This has been the case on Oʻahu in part because of difficulties associated with reaching full agreement on complex issues in this densely populated setting where many active Native Hawaiian interest groups are based. The existence of multiple interest groups and challenging real-time issues also underlay a pattern of forward and stalled momentum on the island of Kauaʻi. "Kinks" in the ʻAha Moku system therefore continue to be ironed out in these distinct island settings, as would be expected given extensive variability in local and regional environments, sociopolitical histories, and other island- and moku-specific attributes both historically and in the present-day.

Despite differences in progress to date, the ʻAha Moku system as a whole has enabled numerous forward strides and accomplishments—in keeping with its mandate to advance traditional natural resource management strategies in ahupuaʻa across the archipelago. For example, as described in formal correspondence between the AMAC and the State of Hawaii House Committee on Water and Land, and in the Committee's 2017 report to the 29th Legislature Regular Session of 2017 (Department of Land and Natural Resources 2017), participants in the ʻAha Moku system continue to advance a core aspect of the overall mission (Kuloloio 2017):

> On behalf of ahupuaʻa communities, the ʻAha Moku [process and Advisory Committee] advises the Chairperson of the Board of Land and Natural Resources (BLNR) on issues pertaining to natural and cultural resources. But more than advising the Land Board, ʻAha Moku has been able to bring ahupuaʻa community concerns to the attention of the divisions of DLNR who then communicate directly with those communities. This continues to progress and is on-going successfully as reflected in the Legislative Report submitted each year. (p. 1)

Fig. 5.4 Cathleen and Tim Bailey working in Haleakalā National Park, Maui in 2012 (Photo courtesy of Elyse Butler Mallams)

As depicted in Table 5.1, many issues and concerns have indeed been brought forward by the AMAC on behalf of residents in various ahupua'a and moku around the islands. These have been, are being, or will be examined for prospective action by the Department of Land and Natural Resources (DLNR) and appropriate agency divisions. While some concern has been expressed that the AMAC has not thoroughly consulted with island-specific standing councils on a consistent basis, again, this may be seen as a kind of growing pain that is subject to correction as the overall process moves forward.

Through years of well-attended meetings, in-depth discussion and deliberation on issues of great importance to local residents, and direct interaction with state agencies and legislative representatives, Native Hawaiians and other kama'āina have "earned" a seat at the natural resource decision-making table in Hawai'i. The many puwalu, 'aha, and individual and collective energies required to convene the meetings and move the 'Aha Moku agenda forward have resulted in a positive outcome for Kānaka Maoli, who now have legal authority to inform and advise natural resource decision-making processes around the islands, and formal state recognition of the 'Aha Moku system of best practices for natural resource management across the islands.

But it must be remembered that many indigenous thinkers, writers, and residents do not in fact recognize the state or federal governments as legitimate entities with authority to determine or otherwise impact the affairs of a sovereign island society, the monarchy of which was illegally overthrown more than a century ago. In this respect, many Native Hawaiians involved in the 'Aha Moku process do not believe anything was "earned," per se, over the years of meetings and deliberation. In fact, many participants originally sought direct decision-making power rather than an advisory role only. Many see the process rather as a temporary form of cooperation with state and federal agencies that are based in the islands, with the much greater attainment of having bolstered a traditional way of life wherein the ahupua'a, and the relationship of the 'ohana to the ahupua'a, are being revivified in a broad and meaningful way. In this respect, the overall natural resource consultation process between residents and representatives of various ahupua'a and moku is proceeding apace irrespective of state and federal agency actions or inactions.

Table 5.1 Natural resource issues and concerns brought to the State of Hawaiʻi Department of Land and Natural Resources through the ʻAha Moku Process[a]

Elicited issues, concerns, and problems	Mokupuni	Responsible DLNR division[b]
Sea Rise and Global Warming	Paeʻāina (all islands)	CO, OCCL
Hawaiʻi State Environmental Court	Paeʻāina	CO, AMAC
Papahānaumokuākea Monument Expansion	Paeʻāina	CO, AMAC
Hunting Issues	Paeʻāina	CO, AMAC, DAR
Marine Spatial Planning, NOAA	Paeʻāina	CO, DOFAW
Community-Based Marine Management, Miloliʻi	Hawaiʻi Island	DAR
Pōhue Bay Illegal Coastal Activities, Kaʻū	Hawaiʻi Island	OCCL
Thirty Meter Telescope Project, Mauna Kea	Hawaiʻi Island	OC, OCCL, SP
Rapid ʻŌhia Death, Puna	Hawaiʻi Island	DOFAW
Keauhou Aquifer, Kona	Hawaiʻi Island	CWRM
Lāʻau Lapaʻau Protocol, Lapakahi	Hawaiʻi Island	SP
Lipoa Point, Kaʻanapali	Maui	SP, DOFAW, DAR, OCCL, SHPD
Honokōhau and Honolua Stream, Kaʻanapali	Maui	CWRM
Wailuku and ʻIao Stream Diversions	Maui	SP, CWRM. SHPD
Small Boat Harbor Improvements, Mānele	Lānaʻi	DOBOR
Ocean Events Management	Molokaʻi	DOBOR, DAR, DOCARE
Bureau of Ocean Management, Ocean Windmills	Oʻahu	CO
Future of Kahoʻolawe	Kahoʻolawe	KIRC
Makua Valley Natural Resource Restoration	Oʻahu	SP, DOCARE
Waiʻanae Water Restoration	Oʻahu	CWRM
Illegal Memorials along Waiʻanae Coast	Oʻahu	DOCARE, SP
Kawainui Marsh Concerns, Koʻolaupoko	Oʻahu	CO, SP
Makai Watch Collaboration	Oʻahu	DOCARE

(continued)

Table 5.1 (continued)

Elicited issues, concerns, and problems	Mokupuni	Responsible DLNR division[b]
Chain of Custody for Sacred Artifacts, Mokulēʻia	Oʻahu	CO, DOFAW, SHPD
Consultation with Taiwanese Resource Harvesters	Oʻahu	AMAC
Ahupuaʻa Stream Protection, Mahaʻulepu	Kauaʻi	CWRM
Nearshore Resource & Fishery Protection, Mahaʻulepu	Kauaʻi	DAR
Ala Loa Trail Restoration, Koʻolau	Kauaʻi	DOFAW—Na Ala Hele
Invasive Rat Eradication, Lehua	Niʻihau	DOFAW—Invasive Species

[a]After DLNR 2017; [b]DLNR Divisions: Chairman's Office (CO), Division of Aquatic Resources (DAR), Commission on Water Resource Management (CWRM), Division of Boating and Ocean Recreation (DOBOR), Division of Conservation and Resource Enforcement (DOCARE), Division of Forestry and Wildlife (DOFAW), Kahoolawe Island Reserve Commission (KIRC), Office of Conservation and Coastal Lands (OCCL), State Historic Preservation Division (SHPD), State Parks (SP)

The outcome to date of the ʻAha Moku effort is truly positive for indigenous Hawaiians—inasmuch as critical aspects of the society and culture are being preserved and advanced while the larger movement toward attainment of the full range of inherent rights continues over time. It may well be that the ʻAha Moku system will play a key role in natural resource management when some form of self-determination is enacted in full (see McGregor and MacKenzie 2014). The perspective that state and federal agencies represent an unwanted layer of governance is not universal across a complex indigenous society, and in fact, many Hawaiians have fought and died for the United States. Yet this is a common and powerful perspective in its extension of resistance to unwanted changes and impacts past, present, and future.

> **Inset E Biography of Mr. Leslie Kuloloio, Chair of the 'Aha Moku Advisory Committee**
>
> A respected kupuna, Les and his 'ohana from Maui have been gathering and fishing on six ahupua'a for more than ten generations. Les believes in a simple Hawaiian lifestyle, which includes fishing and planting, and using one's own natural resources within one's own moku boundaries. The island of Kaho'olawe is part of his ancestral and cultural connections through both of his parents' genealogy. He and his 'ohana are known for taking care of and sustaining the shoreline, reef, and offshore ecosystems of Kaho'olawe and Maui for many years. Les has been productively involved with the Kah'oolawe Conveyance Commission, the Kaho'olawe Island Reserve Commission, and the Protect Kaho'olawe 'Ohana, of which he is the current kupuna spokesperson. Finally, Les sat on the Cultural Resource Management Plans and Land Use Committee prior to the cleanup of ordinances on the island, and participated in marine research of the Kaho'olawe shoreline during a joint venture with the National Oceanic and Atmospheric Administration. (Adapted from 'Aha Moku website, available at http://www.ahamoku.org/index.php/aha-kiole-members/%20)

Constraints and Implications

Unfortunately, the inability of government agencies to consistently fund the 'Aha Moku process has hampered its development. Funds were provided by state and federal entities during initial portions of the effort, but these were expended commensurate with the costs of facilitating and administering such a large initial effort. Some state funding for the process was made available during 2015, and was notable in terms of its source as a partial allotment of settlement funds from a large and impactful molasses spill in Honolulu Harbor (Kaka'ako). This was used to help defray administrative and other costs associated with operation of the AMAC. Appeals for funds needed to effectively administer the 'aha moku process in the present have been denied in recent deliberations by the state legislature.

Significantly, many of the actual costs in time and labor to administer the 'Aha Moku process have been absorbed by the participants themselves, and undoubtedly by the entities that have assisted its development. Despite many achieved and prospective future benefits, the burdens of advancing traditional forms of place-based resource management continue to fall primarily on the indigenous population. This is in part because most public officials and government leaders tend to prioritize matters and fund initiatives deemed more critical or politically expedient vis-à-vis the larger population of residents and visiting tourists.

A fisheries-specific outcome of this situation is that the existing nearshore management regime of statewide gear restrictions, area closures, and other measures remains more static and less sensitive to localized spatial-ecological variability and anthropogenic change than would an ahupua'a-based approach to management of living marine resources. The overall situation is somewhat different in the offshore zone around the islands, where evidence of detrimental local-origin human-environmental impacts is less clear than for the nearshore zone and where tuna and other species in the pelagic zone continue to provide extensive dietary and cultural opportunities for indigenous Hawaiians and other island residents. Although return on effort in the pelagic realm can be significant in economic and sociocultural terms, the funds needed to purchase, maintain, and operate a capable vessel and gear are prohibitive for many.

The fact that the fiscal and human resource costs of advancing traditional forms of resource management are considerable and typically fall on indigenous society is not unexpected in Hawai'i. For instance, Ayers (2016) asserts that communities and their non-governmental partners have absorbed most of the transaction costs associated with movement toward co-management of natural resources. The author states that:

> This goes against theories of institutional change that state resource users will not work to change rules if the costs exceed the benefits. [Residents of] many Hawai'i communities have deep cultural, spiritual, and religious connections to place. This deep kuleana (responsibility or accountability) to care for their area may be why Hawai'i communities have been willing to give up so much in time, effort, and resources with seemingly little in return. (p. 180)

While actual co-management of natural resources—wherein formal government entities equitably share resource kuleana (responsibilities) with resident individuals and entities—is somewhat different than the enhanced representation and locally managed ahupua'a resources engendered by the 'Aha Moku system, the funding outcome is the same: largely insufficient. This also applies to funding needed to better understand and monitor use of natural resources for purposes of subsistence and cultural continuity among Native Hawaiians and other island residents. Indeed, although government agencies in states such as Alaska have for many decades systematically documented the use of natural resources among indigenous and rural residents—thereby contributing to equitable resource management decisions and ideally to the well-being of society as a whole—government agency commitment to undertake such essential work remains entirely lacking in Hawai'i.

Of note, in September 2018, the Department of Land and Natural Resources posted an advertisement seeking an Executive Director for the AMAC—a positive sign that the overall 'Aha Moku process will continue to be administered with a measure of fiscal support from state government. Whether the process can be readily perpetuated in the absence of sufficient funding is uncertain. What is certain, however, is the determination of individuals to carry the momentum of the effort forward in time. Determination, persistence, and the very element of time may be sufficient for success, for as demonstrated over the course of many centuries, Kānaka Maoli are particularly tenacious in their claim to the 'āina and their inherent right to determine the future as a unique and sovereign island society.

5.3   Into the Future

This text has provided a chronologically organized account of the intrepid voyages of early Polynesians from the southern reaches of Oceania to a particularly remote archipelago in the Central Pacific now called the Hawaiian Islands. The mo'olelo (story) of this journey, subsequent phases of colonization and population growth, and centuries of selective adaptation and resistance to change, is precious to descendants of the original colonists. Past, present, Kānaka Maoli, 'ohana, island, ocean, and culture are indivisible in the hearts and minds of many to this day. Expressed in the ancient creation chant known as the

Kumulipo, persons unfamiliar with Hawai'i and Hawaiians will have to suspend their own ethnocentric perspectives to comprehend this profound system of understanding and action.

This conceptual and practical system of interaction with the natural world, coupled with accumulated and evolving knowledge of its ecological subtleties and variability, form the basis of tradition to which indigenous people of the islands turn for food, guidance, identity, and a sense of individual and collective wholeness in the present. Large-scale efforts and movements to advance this body of traditional knowledge and experience into the future, such as the *Ho'ohanohano I Nā Kūpuna Puwalu* series and subsequent assemblies, are both challenging and invaluable in a contemporary setting of rapid technological, demographic, and social change. Given the brief nature of human life, and the limited capacity of individuals to effect positive change in the contemporary island setting of competing perspectives and ideologies, such group efforts are essential for transmitting social and cultural objectives across generations. Native Hawaiians may well be best practiced and

Fig. 5.5 Tradewind showers generate a vivid rainbow at Hale'iwa Harbor on O'ahu

prepared to meet this need, for ʻohana past and present, ʻōpio (youth), and generations yet to come, remain at the heart of the culture.

Kūpuna (elders) have long retained and communicated practical knowledge that derives from the past, addresses action in the present, and is to be applied in the future. The following passage from Maly and Maly (2003, Volume II: 1255) provides useful and fitting words of wisdom and parting advice from Elia Kāwika Kapahulehua, the first Hawaiian to captain an ocean-voyaging canoe from Hawaiʻi to Tahiti in modern times. Born on Niʻihau in 1930, Elia has now passed on from his treasured ʻohana and island home (Fig. 5.5):

> Our ancestors said "Mai ʻuhaʻuha" (Don't be greedy)! Because the ocean is our ice box. You take what you need for today, you come back tomorrow. There is still some for tomorrow, or the next day, or the day after, and next week. So, take what you need just for the day. But sometimes, when you throw the net, you get more fish in the net, you think of your neighbors, share with them, help them. Tell them, "I cannot help it. Aʻole hiki ke ʻalo aʻe, kiloi wau i ka ʻupena, paʻa mai iaʻu kaiʻa!' I caught all this fish, and ka wehewehe ʻana, as I take it from the net, all make (dead)." So I bring it for you, to share with you. So mai ʻuhaʻuha, lawe ka iʻa i ʻai ʻoe i kēia lā, kou pahu hau kēlā. That is, the ocean is your refrigerator. So take care of it. Teach the children to know that there is tomorrow.

References

Ayers, A. (2016, December). *From Planning to Practice: Toward Co-Management of Hawaii Coral Reef Fisheries*. A dissertation submitted to the Graduate Division of the University of Hawaii at Mānoa in partial fulfillment of the requirements for the degree of Doctor of Philosophy in Urban and Regional Planning. Honolulu.

Cluett, C. (2012, March 12, Sunday). Passenger Boats Survey Results. *The Molokai Dispatch*.

Hawaii News Now. (2013, January 9). Cruising to Compromise on Molokai.

Heen, W. M. (2008). Taking a Seat at the Table. Puwalu ʻEkolu (Third Conference): Opening Remarks (Day 2). In *Current—The Journal of Marine Education, 24*(2). Special Issue featuring *Hoʻohanohano I Nā Kupuna*

Puwalu and International Pacific Marine Educators Conference. K. Simonds, S. Spalding, and C. Kaaiai, Western Pacific Regional Fishery Management Council (Issue Editors). Available at http://www.wpcouncil.org/wp-content/uploads/2013/04/GCRL-145865_finalfinal.pdf.

Hui Mālama O Moʻomomi. (2017). *Moʻomomi—North Coast of Molokaʻi. Community-Based Subsistence Fishing Area Proposal and Management Plan.* Submitted by Hui Mālama O Moʻomomi on Behalf of the Hoʻolehua Hawaiian Homesteaders to the State of Hawaiʻi Department of Land and Natural Resources, Divison of Aquatic Resources. Honolulu.

Kuloloio, L. (2017, February 8). *Testimony of Leslie Kuloloio Before the State of Hawaiʻi House Committee on Water and Land in Support of House Bill 231 Relating to the ʻAha Moku System; ʻAha Moku Advisory Committee; Funding.* Honolulu.

Liliʻuokalani Trust. (2018). *Community Profile: Molokaʻi. Division of Research and Evaluation.* Honolulu. Available at https://onipaa.org/pages/research-and-evaluation.

Maly, K., & Maly, O. (2003). *Ka Hana Lawaiʻa A Me Na Koʻa O Na Kai ʻEwalu: A History of Fishing Practices and Marine Fisheries of the Hawaiian Islands* (Vols. I and II). Hilo: Kumu Pono Associates. Prepared for the Nature Conservancy and Kamehameha Schools.

Matsuoka, J., McGregor, D., Minerbi, L., & Akutagawa, M. (Eds.). (1994). *Governor's Molokaʻi Subsistence Task Force Final Report.* State of Hawaiʻi. Honolulu: The Task Force.

McGregor, M., & MacKenzie, M. (2014, August 19). *Moolelo Ea o Na Hawaii—History of Native Hawaiian Governance in Hawaii.* Report prepared for the Office of Hawaiian Affairs. Honolulu.

Poepoe, K. K., Bartram, P. K., & Friedlander, A. M. (2003). The Use of Traditional Hawaiian Knowledge in the Contemporary Management of Marine Resources. In *Putting Fishers' Knowledge to Work* (pp. 328–339), Fisheries Centre Research Report, University of British Columbia. Vancouver, Canada.

Poepoe, K. K., Bartram, P. K., & Friedlander, A. M. (2007). The Use of Traditional Knowledge in the Contemporary Management of a Hawaiian Community's Marine Resources. In N. Haggan, B. Neis, & I. G. Baird (Eds.). *Fishers' Knowledge in Fisheries Science and Management.* Coastal Management Sourcebooks 4. Paris: United Nations Educational, Scientific, and Cultural Organization (UNESCO).

State of Hawaiʻi. (2012, July 9). 26th Legislature. Act 288 Relating to Native Hawaiians.

State of Hawai'i. (2016). *Final Rules of Practice and Procedure Department of Land and Natural Resources 'Aha Moku Advisory Committee.* Available at http://www.ahamoku.org/wp-content/uploads/2016/12/FINAL.AMAC_. Admin_.Rules_.effective.102016.pdf.

William S. Richardson School of Law. (2018). *Personnel Biography: Malia Akutagawa '97.* Faculty and Staff Directory, Biographical Profiles. University of Hawaii at Manoa. Honolulu. Available at https://www.law.hawaii.edu/personnel/akutagawa/malia.

Glossary for Hawaiian and Other Polynesian Terms

Pronunciation Hawaiian vowels are as in English: a, e, i, o, and u. But with respect to pronunciation, the letter "a" is pronounced as the soft *ah* sound in papa; "e" as the ā sound in play; "i" as the ē sound in need; "o" as used in bowl; and "u" as the ew sound in tune. Diacritical marks are used to indicate stress on particular vowels, and as glottal stops. The macron (called kahakō in Hawaiian) is used to stress and elongate any of the vowel sounds. For example, the ā sound in pāhoehoe (sheet lava) is stressed and lengthened, as in p*ahh*-ho-ay-ho-ay. The reverse apostrophe (called an okina in Hawaiian) is used as a glottal stop, as in the closed throat sound that should precede formation of the word 'ahi (pronounced *ah*-hee), or between sounds, as in Punalu'u (pronounced poo-nah-lew-ew). Certain vowel combinations (diphthongs) are also pronounced in a manner dissimilar to the way they are pronounced in English, with stress on the first vowel. For instance, the "ou" sound in Hawaiian is pronounced with stress on the o, as in pouli

E. W. Glazier, *Tradition-Based Natural Resource Management*,
Palgrave Studies in Natural Resource Management,
https://doi.org/10.1007/978-3-030-14842-3

(Hawaiian for dark or eclipse, pronounced poh-lee). Pronunciation of consonants is as in English, although the w is often pronounced as an unstressed v, as in Kahoʻolawe, pronounced kah-ho-oh-lah-vay).

Sources The following Hawaiian spellings and definitions derive primarily from the *Hawaiian Dictionary* (Pukui and Elbert 1986), Ka puke wehewehe a Pukui/Elbert (2018), and Andrews and Silva (2003). Definitions of certain terms used to describe fishing methods derive from Nogelmeier (2006), as cited in the glossary.

*   *   *

aʻa niu	coconut cloth
ʻaha	meeting, assembly, convention
ʻaha moku	standing councils of representatives from the various island districts
ʻaha kiole	council initially overseeing ʻaha moku
ʻahi	Yellowfin tuna (*Thunnus albacares*)
aho	fishing line, cord, lashing, trolling line
aho kākele	trolling line; also aho kālewa
aho loa	long line, as with several hooks for deep-sea fishing or sounding
ahu	shrine comprised of piled stones
ʻahuluhulu	juvenile kūmū (white saddle goatfish; *Parupeneus porphyreus*)
ahupuaʻa	land division within an island district, usually extending from the uplands to the sea
ʻai kapu	prohibitions concerning eating
ʻai noa	release from eating prohibitions
akamai	savvy, knowledgeable, smart
aku	Skipjack tuna (*Katsuwonus pelamis*); pelagic food fish, historically popular among Hawaiians
akua	god, goddess, spirit
akule	bigeye scad (fish); *Selar crumenophthalmus*
ala hele	path, route, road
aliʻi	chief, chiefess, divine class, etc.

ali'i nui	high-ranking chief
'āina	earth, land
'alalaua	Hawaiian bigeye (fish); *Priacanthus alalaua*
'ama'ama	mullet fish (*Mugil cephalus*)
ana	measure, evaluate
'anae	full-sized 'ama'ama
Aotearoa	New Zealand
aouli	firmament, sky, blue vault of heaven
apoālewa	the space above, where the birds soar
a'u	swordfish, sailfish, marlin, spearfish
a'ua'u	small swordfish, sailfish, marlin, or spear-fish; pelagic; historically popular food fish in Hawai'i
'auhau	tribute
'aumakua	family god, typically assuming the form of an animal, such as shark, owl, hawk, octopus, and so forth
aupuni	government
'awa	kava (*Piper methysticum*)
'aweoweo	Hawaiian bigeye (fish); *Priacanthus meeki*
eaea	air
ehu	important deep water food fish—Squirrelfish snapper; *Etelis carbunculus*
'ekahi	one, counting in a series
'ewalu	eight, eight times
Fa'a Samoa	traditional Samoan lifeways
hāhā	method for capturing fish with one's hands, also a trap made of twigs and small branches, for fresh-water fish
hahai	hunt
haku	lord, master, overseer
hala	*Pandanus* spp., tree native to Hawai'i
hālau	as used here, a school—as for hula instruction
hāloa	poetic name for lauloa taro; far-reaching
hānai	to raise, rear, feed, nourish, sustain, as in adopting a child

hanauna	generation, generations
haole	foreign, foreigner, Caucasian
Hawai'i nei	this Hawai'i
he'e	squid (*Euprymna scolopes*)
hei'au	shrine, temple
hī	to cast or troll, as for 'ahi or kala
hīnālea	small to moderate-sized brightly colored wrasse (fish); family *Labridae*
hō'ale	to form waves, stir
hoa'āina	native inhabitants
hō'au'āu	to set a net, such as an 'upena ku'u (lay net)
hō'ihi	treat with respect
holoholo	net into which fish swim after being frightened; to fish with this net; sometimes used in conjunction with a weir
holo'oko'a	whole, entirety
holomua	progress, improvement
ho'o holo	to cause to move or run, as in a fish that is nudged to swim
ho'okupu	offerings given as tribute
ho'olu'ulu'u	small rounded basket trap for hīnālea, historically made of vines and sticks
ho'o mau	to continue, persist, persevere, last
ho'omo	kind of canoe used especially for trolling aku
ho'omoemoe	type of nighttime fishing with net ~60 fathoms in length with mesh two fingers wide
hō'oni	to cause to move
ho'onohonoho ku'una	traditional management
ho'oponopono	to make right
hukilau	seine nets, to fish with such nets
hula	form of expressive dance developed by indigenous Hawaiians
huna	secret
i'a	fish, marine animal
'iako	outrigger boom

'iao	bait fish—silversides (Hawaiian silverside; *Atherinormorus insularum*)
iheihe	halfbeak (fish—*Hyporhamphus pacificus*)
'ike	knowledge
'ike ponolia	practical knowledge
'ili	subdivision of an ahupua'a, often used by specific 'ohana for generations
'ilihahi	sandalwood trees, endemic to Hawai'i
'imi fenua	searching for lands (in Marquesan)
'imi honua	searching for lands (in Hawaiian)
iwi	bone, bones
ka'ā	thread, line, as of olonā fiber; leader or snell of a fishline, snood; ply, twist, strand; to make thread, as of olonā
kahiko	old, ancient, antique
kahuna	priest, sorcerer, expert in any profession (plural kāhuna)
kāhuna lapa'au	Hawaiian medical specialists
kāhuna lawa'ia	fishing experts
kai	sea
kai leo nui	loud sea
ka'i	fish net or seine; snare or noose for birds
kaiāulu	community
kāili	to cast for fish
kā'ili	string of fish, string or fiber on which fish are strung
kaikamahine	daughter
kala	surgeonfish (*Acanthurus* spp.), unicorn fish (*Naso* spp.)
kā lā'au	a kind of fishing wherein the water was beaten with sticks to scare fish such as 'ama'ama into pāloa (long seine-like surround nets (from Nogelemeier 2006))
kala ku	net fishing for kala with mesh two to three fingers wide for schooling fish such as kala (from Nogelmeier 2006)

kalana	division of land smaller than a moku or district and bigger than an ahupua'a
kalekale	Von Siebold's snapper; *Pristipomoides sieboldii*
kalo	taro or *Colocasia esculenta*
kama'āina	(vernacular)—one who has lived in Hawai'i for a long time
kanaka paeaea	pole fisherman
Kānaka Maoli	the original colonists of the Hawaiian Islands and their descendants (plural)
Kānaka 'ōiwi	Native sons of Hawai'i
kānāwai	law, rule, code
Kāne	one of the four great Hawaiian gods: Kāne, Kū, Lono, and Kanaloa
kane	man
kaona	literary device; hidden or double meaning
kapa	tapa, bark cloth
ka po'e kahiko	people of old
kapu	stricture, prohibition, forbidden, taboo
ke	definite article, often translated as "the"
keiki	children
kī	ti (*Cordyline terminalis*), as in lau kī (ti leaf)
kīhāpai	small land division, cultivated patch, garden, orchard, field, small farm
ki'i pōhaku	petroglyph
kiolauola	a kind of fishing from the shoreline or wharf, similar to ku'iku'i
kipuka	geographic regions or pockets where Hawaiian cultural traditions of old continue to evolve
koa	*Acacia koa*; the largest of Hawaii's native forest trees; a highly valuable lumber tree, used for canoes, surfboards, calabashes, furniture, and ukuleles
ko'a	fishing ground
ko'a huna	secret fishing ground
kō'ele	small land unit farmed by a tenant for the chief
kōkua	help, aid, assistance, cooperation

kolo	fishing method involving use of a large bag-like net
konohiki	overseer of land for a chief (the haku 'āina)
ku'ai	barter
kua'āina	back country, person from the back country
kuapā	walls of a fish pond
ku'iku'i puhi	to pound eel for bait for ulua
kūkaula	handline fishing in the deep sea
kuleana	right, privilege, concern, responsibility
kumu	source, foundation, basis; teacher, tutor
kūmū	white saddle goatfish (*Parupeneus porphyreus*)
kumukānāwai	code of law
kumulau	roots with many sprouts
Kumulipo	sacred Hawaiian cosmological genealogy
kūpapakū	bedrock
kūpuna	elders, ancestors
kupuna kāne	grandfather
kupuna wahine	grandmother
ku'una	traditional
ku'una	to let down a fishing net, place where a net is set in the ocean
la'a	sacred
lā'au	plant, such as kalo
lāhui	Hawaiian nation, people, tribe
lae'ula	a well-trained, clever person; an expert
lamakū	torch
lani	sky, heavens
lau	leaf
laulima	literally many hands, cooperation, joint action to achieve a collective goal
lauloa	variety of long-stalked kalo or large-stalked sugar cane; also long wave or surf, as extending from one end of the beach to the other
lawai'a	fishing
lawai'a 'ili'ili	fishing assistants, as used by Kamakau (1976: 4–5)

lawelawe	work, service, function
lawelawe hana	administration
lawena	acquiring, acquisition
lepo	ground, earth
leo	voice, tone, melody
limu	general name for all kinds of plants living under water and along the shoreline, especially seaweeds
lohe	listen, heed, obey
lo'i	irrigated terrace
lōkahi	unity
loko	pond, lake, pool
loko i'a	fish pond
loko i'a kalo	ponds in which fish and kalo were grown together
lū'au	feast
luelue	bag net with mesh size one finger wide; held open by a hoop and baited and lowered into the sea by long cords
luna	foreman, boss, overseer
lunamaka'āinana	representative
mahele	apportionment, land division of 1848
mahi'ai'ana	farming
mahimahi	dolphinfish (*Coryphaena lippurus*); popular food fish in Hawai'i and elsewhere; pelagic
mai'a	plantains and bananas (*Musa* spp.)
ma'i 'ōku'u	Hawaiian term for early epidemic causing severe loss of hydration—possibly cholera, typhoid, or bubonic plague
maka'āinana	commoner, citizen
mākāhā	sluice gate, as of a fish pond; entrance to an enclosure
makai	toward the ocean
Makahiki	ancient festival beginning about the middle of October and lasting about four months, with sports and religious festivities and kapu on war

makau	fishhook
makua	parents, and relatives of one's parents
makua kāne	father
māla	garden, plantation, patch, cultivated field
mālama	care, tend
malau	bait carrier
mālolo	flying fish (*Parexocroetus brachypterus*)
manawa	time, a time
maomao	a damselfish, favored by ali'i for its tenderness
mana	divine energy or power
manō	general name for sharks, of which there are many kinds in waters surrounding Hawai'i and other part of the Central and Western Pacific
mauka	inland, toward the mountains
marae	stone temple, religious statuary, sacred place (Tahitian)
mele	poem, song, chant
Mele Kalikimaka	Merry Christmas
melomelo	club smeared with bait to attract fish to a net
moana	ocean
mōhai	sacrificial offering
moi	Pacific threadfin (fish); *Polydactylus sexfilis*
mō'ī	king, monarch, queen
mokuna	chapter
molekumu	root, origin
mo'olelo	story, tale, history, tradition
mo'okū'auhau	genealogy
mo'omeheu	culture
moku	island district
mokupuni	island
na'auao	knowledge, wisdom, science
nae puni	fine mesh nets supported by sticks
nānā	look, observe
nehu	anchovy (*Stolephorus purpureus*); for consumption and bonito palu

nenue	rudderfish; *Kyphosus spp.*
noa	freed of taboo, released from restriction
nui	large
nupepa	newspaper
ʻō	spearing method for capturing fish
ʻohana holoʻokoʻa	extended family
ōʻio	bonefish (*Albula volpes*)
ʻokana	land district or subdistrict, usually comprising several ahupuaʻa
ʻōkilo	to observe, watch carefully
ʻōkuʻu	dysenteric disease at time of Kamehameha I; lit., to squat
ʻōlelo paʻiʻai	Hawaiian Creole English, *lingua franca* used on Hawaiʻi plantations
oli	Hawaiian chant
olonā	native shrub of Hawaiʻi (*Touchardia latifolia*), used for cordage and nets
ono	important pelagic food fish; wahoo or *Acanthocybium solandri*
oʻo	mature, ripe, potent
oʻopu	General name for fishes included in the families *Eleotridae*, *Gobiidae*, and *Blennidae*. Some are in salt water near the shore, others in fresh water, and some said to be in either fresh or salt water
ʻōpakapaka	deep water snapper species; esp. Pink snapper (*Pristipomoides filamentosus*)
ʻopelu	mackerel scad (*Decapturus spp.*)
ʻopihi	limpets (*Cellana* spp.); a Hawaiian delicacy
ʻōpio	young persons, youth
pā	mother-of-pearl shell fishhook
paʻakai	salt
paeaea	to fish with a light pole; pole fishing
paeʻāina	archipelago
pāhoe	type of canoe-based ulua fishing wherein the bait and hook are trailed along behind with a bundle of mashed food to attract fish

pā'ina	meal, party with dinner
pākehā	Māori term for persons of European descent
palapala	documentation, writing of any kind
palu	mashed or softened bait used to attract fish, chum
palu 'ahi	ancient tuna handline fishing method
Papa	earth mother
pāpa'i	general name for crabs
Papakōlea	Native Hawaiian homestead land and neighborhood above Punchbowl in Honolulu
papālagi	Samoan term for non-Samoan, white person
pekeu	wing, as of a bird or fish such as 'ahi
pīhā	Round herring (*Spratelloides delicatulus*); <8 cm long; historically preferred dried and salted
pō	darkness, twenty-four hour period
pō'alima	obligations to work on the chief's plantation, done on Fridays
po'e	people, persons, assembly, assemblage
po'e lawa'ia	fishermen
poi	starchy paste made from cooked taro corms pounded & thinned with water
pono	good, correct
popa'a	Tahitian term for non-Tahitian, European, or white person
pōpolo	Black nightshade (*Solanum nigrum*); same as maiko, surgeonfish (*Acanthurus nigroris*) (Ni'ihau)
pouli	dark, eclipse
pououo	bag net with mesh two fingers wide, similar but larger than the luelue
pua'a	pig
puhi ki'i	juvenile flying fish or mālolo
pūlama	cherish or care for
pule	prayer, prayers, incantation, blessing, grace, magic spell
pu'u	in terms of geology: steep hill, cinder cone, peak

pu'u one	sand dune banks or walls, as used for shoreline fish ponds
Puwalu	people working together, in unison, united, cooperative
Rapa Nui	Easter Island
tūtū	kupuna
'uala	sweet potato (*Ipomoea batatas*)
uhane kia i	guardian spirit
uhu	parrotfish (*Scarus spp.*); common edible reef fish
uka	inland, upland, towards the mountains
'ulu	breadfruit (*Artocarpus altilis*)
ulua	various species of trevally (fish), in Hawai'i especially *Carynx ignobilis* (Giant trevally)
uouoa	false mullet or false 'ama'ama; (*Neomyxus chaptalii*); historically, the head, when eaten, was said to cause sleeplessness and nightmares
'upena	fishing net
'upena'apo'apo	gill net
wa'a	outrigger canoe
wa'a pā 'ekolu	three-board canoe
wahine	woman
wai	fresh water
waiwai	resources, assets, goods
Wākea	sky father, from whom islands are born
wana	urchin
weke	goatfish (*Mulloidicthys spp.*); in historic times, weke were used as offerings to the gods to turn away curses; also associated with nightmares

Proverbs

I ka wā ma mua, ka wā ma hope
In the past, the future

I ulu no ka lala i ke kumu
The branches grow because of the trunk; (figuratively), without our ancestors we would not be here.

Ka lamakū o ka naʻauao
The torch of wisdom

References

Andrews, L., & Silva, N. (2003). *A Dictionary of the Hawaiian Language.* Waipahu, Hawaiʻi: Island Heritage Publishers.

Kamakau, S. M. (1976). *Na Hana a ka Poʻe Kahiko* (The Works of the People of Old). Translated from the Newspaper Ke Au ʻOkoʻa by Mary Kawena Pukui. Arranged and edited by D. Barrere. Bernice Bishop Museum Special Publication 61. Bishop Museum Press. Honolulu.

Na Puke Wehewehe ʻŌlelo Hawaiʻi. (2018). Hawaiian Electronic Library. Copyright owners: Ka Haka ʻUla O Keʻelikolani College of Hawaiian Language, and Alu Like, Inc. Available here.

Nogelmeier, P. (2006). *Ka ʻOihana Lawaiʻa – Hawaiian Fishing Traditions.* M. Puakea Nogelmeier (ed.). Translations by M. K. Pukui. Bishop Museum Press. Honolulu.

Pukui, M. K., & Elbert, S. H. (1986). *Hawaiian Dictionary* (6th ed.). Honolulu: University of Hawaii Press.

Index

The manufacturer's authorised representative in the EU is Springer
Nature Customer Service Centre GmbH, Europaplatz 3, 69115 Heidelberg,
Germany. If you have any concerns regarding our products, please
contact ProductSafety@springernature.com

Printed and bound by CPI Group (UK) Ltd, Croydon, CR0 4YY

29/04/2026

02099471-0007